Null Hypothesis Testing:

Demystified!

Null Hypothesis Testing:

Demystified!

Frank S. Corotto

Published by Frank S. Corotto

ISBN-13: 978-1086162561

Table of Contents

What Makes This Book Different? Please Read This Preface

Few students sitting in their freshmen-level *Probability and Statistics* class learn that the practice of testing null hypotheses has been the subject of fierce debate for decades. Criticisms include the assertions that the null can never be correct, that null hypothesis testing is grossly misunderstood, that testing nulls leads to mindless accept/reject decisions, and that focus on the results of null hypothesis tests comes at the expense of other aspects of data. One purpose of this book is present null hypothesis testing in a form that is immune to those criticisms—to present null hypothesis testing as it should be understood and carried out.

Traditionally, null hypothesis tests have been performed in a mindless, step-by-step fashion, which Gigerenzer[1] calls the "null ritual" and Salsberg[2] describes as "exact", "didactic", and "rigid".[3] Remarkably, no one dreamed up this mechanical approach. It evolved on its own from two schools of thought. The first is generally attributed to Ronald Fisher. Though he did not create null hypothesis testing, Fisher coined *null hypothesis* and strongly advocated for a certain approach to testing nulls. Fisher's hated colleagues, Jerzy Neyman and Egan Pearson, tried to embellish upon Fisher's method but, instead, they created a different kind of test best suited for different circumstances.[4] Today, null hypothesis testing is often carried out as a weird hybrid of those two schools of thought—Fisher's on the one hand and Neyman and Pearson's on the other—a hybrid with flaws.

Many of those flaws can be addressed by using most of Fisher's approach and dispensing with much of Neyman and Pearson's. Their method encourages mechanical decision making based on probabilities. Fisher's method is more streamlined and encourages us to think more about our data, not less. One thing that makes this book different is its strong, Fisherian leaning.

---

[1] G. Gigerenzer. 2004. The Journal of Socio-Economics, 33: 587–606.

[2] D. Salsburg. 2001. The Lady Tasting Tea. Holt, p. 109.

[3] I follow the Council of Science Editors' recommendations when it comes to quotation marks.

[4] e.g. D. Perezgonzalez. 2015. Frontiers in Psychology, 6: article 223.

Another thing that makes this book different is that there is little emphasis on *reject* or *fail to reject* decisions. The reason is that, quite often, the null hypothesis cannot be correct in the first place. In 1991, John Tukey explained why this fact does not matter. We test nulls not to see if they are true or false, but to see if we can rule out sampling error as the cause of a difference in a certain direction.[1] *A* is greater than *B* in our samples. Can we say that *A* is greater than *B* in general, or could it be the other way around? Tukey *should have* changed the way we think about null hypothesis tests, but the textbooks never changed, and countless persons have gone on failing to reject—or worse, accepting—null hypotheses that cannot possibly be true.

So, I present most of Fisher, a little of Neyman and Pearson, and a key contribution from John Tukey. I hope the result is a coherent, practical method of inference. Null hypothesis tests only tells us about single sets of data. They play a limited role, but an indispensable role, among other methods of inference.

[1] J. Tukey. 1991. Statistical Science, 6: 100–116.

CHAPTER 1
THE POINT IS TO GENERALIZE BEYOND THE RESULTS

1.1 Samples and Populations

Suppose we work for a company that manufactures mouse feed. We develop a new feed, Brand X, and wonder if mice prefer it to Brand W, our best-seller. We randomly choose ten CF-1 mice from our colony and use Ivlev's forage index to determine each mouse's preference. If a mouse prefers Brand X, forage index is greater than zero. Negative forage indexes would be generated if a mouse preferred Brand W. The result is a statistical sample (or simply *sample*), ten forage indexes, one from each mouse. A statistical sample should not be confused with the act of collecting data, i.e., sampling. A statistical sample is a set of numbers. Within a sample, each value—each mouse's preference—can be called a **datum**, an **observation**, or a **replicate**. In statistics, **replication** does not refer to repeating an entire study. It refers to having more than one datum in a sample.

Our goal is to generalize beyond the ten mice to all the CF-1 mice in the colony. The preferences of all those mice make up the **statistical population**, the larger set of numbers that we want to know about. It is usually not feasible to collect data from an entire population, so we use samples to draw inferences about populations, the practice of **inferential statistics**. In contrast, if we determined the preference every mouse in the colony, we could calculate the population mean and be done. We would be practicing descriptive statistics.

Some equate samples with the subjects (or *units*), in this case mice, but there is a tradition of calling a set of data a sample. Here, I follow Zar,[1] and consider both samples and populations to consist of numbers.

---

[1] J. Zar. 2010. Biostatistical Analysis, 5th ed. Prentice Hall. pp. 16,17.

1.2 Real and Hypothetical Populations

Note that, in the example of mouse preference (Section 1.1), both the sample and population would be **real**. Existing mice would have existing preferences for one food or another. The situation would differ if we wanted to know the effect of 5% ethyl alcohol on daphnias,[1] which are small crustaceans. We might randomly choose ten daphnias from a container, expose them to alcohol, and determine their heart rates. Those heart rates would make up the sample, but what is the population? It would be the heart rates of all the daphnias in the container were those daphnias to be exposed to 5% alcohol. The population would be **hypothetical**. Populations are nearly always hypothetical in **experiments**, which are cases in which we manipulate an independent variable and study the results.[2]

1.3 Randomization

In scientific studies, there is random sampling and random assignment. In **random sampling**, a specific procedure is used to ensure that all subjects have an equal probability of being chosen for study. If we are choosing CF-1 mice from our colony, we could assign every mouse a name, enter those names into a spreadsheet, use the spreadsheet to assign a random number to each name, sort the names according to those numbers, and choose the ten mice at the top of the list. Thanks to random sampling, a sample is as representative of a population as possible, given the inevitable sampling error (Section 2.1).

In **random assignment**, all subjects have the same probability of being assigned to the various comparison groups. Suppose we want to know the effect of alcohol on the heart rate of daphnias. After randomly choosing 20 daphnias to study, we could use a specific procedure to ensure that each one has an equal probability of being immersed in either a solution containing alcohol or a control solution. The point of random assignment is to limit the effects of confounds.[3,4]

Suppose we randomly assign daphnias to either *alcohol* or *control,* and wind up with 14 in the former and six in the latter. We would prefer equal sample sizes. To achieve that, we could use **randomization without replacement**. We could cut paper into 20 pieces of equal dimensions, write *A* on 10 of them and *C* on the rest, place the squares

---

[1] The genus is *Daphnia*, but there is no plural form for generic names.

[2] W. Shadish, T. Cook, and D. Campbell. 2002. Experimental and Quasi-experimental Designs for Generalized Causal Inference. Houghton Mifflin Company. p. 507.

[3] For a more technical explanation of randomization's benefit, see W. Shadish, T. Cook, and D. Campbell. 2002. Experimental and Quasi-experimental Designs for Generalized Causal Inference. Houghton Mifflin Company. pp. 248–251.

[4] Fisher was a strong proponent of randomization.

in a hat and, for each daphnia, draw one square. Each time we do so, we do not put the square back in the hat. We do not *replace* it. The last square we draw would not be drawn at random—it would be the only square left—but that is how randomization without replacement works.

It is important to distinguish between doing something randomly and doing it haphazardly. If we simply pluck 20 daphnias from a jar, it would not be random. It would be **haphazard**. We are most likely to choose the big ones. They will have less surface area to volume, and will absorb alcohol more slowly, than the daphnias in jar in general. We could underestimate the effect of alcohol. Similarly, if we assign them to one group or the other haphazardly, we might favor putting the smaller daphnias in alcohol, to maximize its effect. Random assignment would eliminate that "selection bias".[1]

1.4 Know Your Population, and Do Not Generalize Beyond It

When using samples to draw inferences about populations, it is essential to define the population correctly. A **population** may be thought of as ***that which is sampled***. In the case of mouse preference (Sections 1.1 and 1.2), the population is not the preferences of all mice, nor is it the preferences of all CF-1 mice.[2] It is the preferences of all CF-1 mice in the colony. In applying inferential statistics, it is important not to generalize beyond the population. In this case, we can only draw inferences about the CF-1 mice in our colony.

Suppose we study CF-1 mice in our colony, decide they prefer Brand X, and consider that to be the preference of all CF-1 mice in the world. We would be guilty of pseudoreplication. The subjects, or units, are no longer the mice. There would be just one subject: the colony. That colony would only generate one valid replicate or datum, such as the mean forage index. In **pseudoreplication**, the individual data are tied together, but they are treated as if they were independent of each other. The data come from one colony, so that is what ties them together. If we only generalize our finding to the colony, the data are true replicates, and our conclusion is valid. If we generalize beyond, the data are pseudoreplicates, and our sample size would be one—the mean forage index of the mice chosen for study. We cannot draw a conclusion about CF-1 mice in general when the sample is size is one.

Suppose we only want to determine the preference of CF-1 mice in our colony, but we house the mice six to a cage. The subjects, or units, are no longer the mice. The

[1] I follow the Council of Science Editors' style manual when it comes to quotation marks.

[2] CF-1 mice are inbred to the point of genetic uniformity. It should be safe to generalize to all CF-1 mice but, technically, we should not do so.

subjects are the cages. Everything is fine, provided we randomly choose one mouse from each cage. If we choose two, their preferences would have to be averaged, and that average would be the replicate.

How do we avoid pseudoreplication? Often, by mixing things up. Imagine that mice in one room receive a treatment and mice in another serve as controls. The subjects are the rooms. The solution is to mix up the mice, so that both experimental and control mice are housed in the same room or in both rooms.

Pseudoreplication is such an important issue that it is important to recognize Stuart Hurlbert for coining the term and emphasizing its threat. Hurlbert's classic *Pseudoreplication and the Design of Ecological Field Experiments*[1] should be closely studied by everyone who reads this book. The examples in this chapter are obvious, but pseudoreplication can be subtle.

[1] S. Hurlbert. 1984. Ecological Monographs, 54: 187–211.

CHAPTER 2
NULL HYPOTHESIS TESTING EXPLAINED

2.1 Why We Need to Test Null Hypotheses

Continuing with the example of mouse preference (Chapter 1), imagine we obtain the following sample of forage indexes from the CF-1 mice we randomly choose from our colony.

	Forage index
	0.12
	-0.09
	0.14
	-0.08
	0.31
	0.10
	0.76
	0.44
	-0.03
	0.12
mean ($\bar{x}$) =	0.19

Example 2.1 A sample of forage indexes.

Since the average ($\bar{x}$) is positive, it appears that the mice prefer Brand X, but that is the average of the sample. We need to generalize to the population, the preferences of all the CF-1 mice in the colony. Just because the mice we randomly chose prefer Brand X does not mean that the same is true on average for all the CF-1 mice in the colony. In spite of random sampling, there is still chance, which creates **sampling error**: a sample is never a perfect representation of a population. To illustrate, imagine that the mean forage index of the CF-1 mice in our colony is zero, no preference, and we collect ten random samples. Every sample mean will be different, and none of them will be zero to an infinite number of decimal places (Figure 2.1). **A sample mean is only an estimate of a population mean**. So, the mean forage index of 0.19 that we obtained may be due to sampling error alone, rather than a preference of the colony.

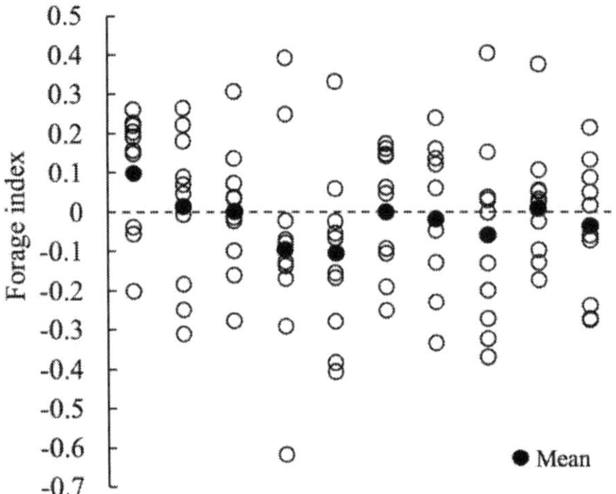

Figure 2.1 Ten random samples from a normally-distributed population with a mean of zero.

The **purpose of null hypothesis testing** is to determine if we can exclude sampling error as the sole reason for a particular outcome, such as 0.19. That outcome is often a difference in a certain direction, a **directional difference**. Our sample mean differs from zero by 0.19, and the difference is in the positive direction.

2.2 The Logic of Testing a Null Hypothesis

To understand null hypothesis testing, it is first necessary to understand the hypothetico-deductive method. We begin with a hypothesis—The CF-1 mice in our colony prefer Brand X over Brand W (Chapter 1). Like any hypothesis, this one exhibits the following features:

- It is a broad statement about the population. It concerns the preferences of all the CF-1 mice in our colony.
- It is a firm statement. This is the way that it is. It does not sound hypothetical.
- It is in the present tense.

With that hypothesis in hand, we next formulate a prediction of what we should find if that hypothesis were true (Figure 2.2A). If we randomly choose ten CF-1 mice from our colony, and determine each one's forage index, we predict that the mean forage index will be greater than zero. The prediction has the following characteristics.

- It is a narrow statement about the data we will collect.
- It is in the future tense.

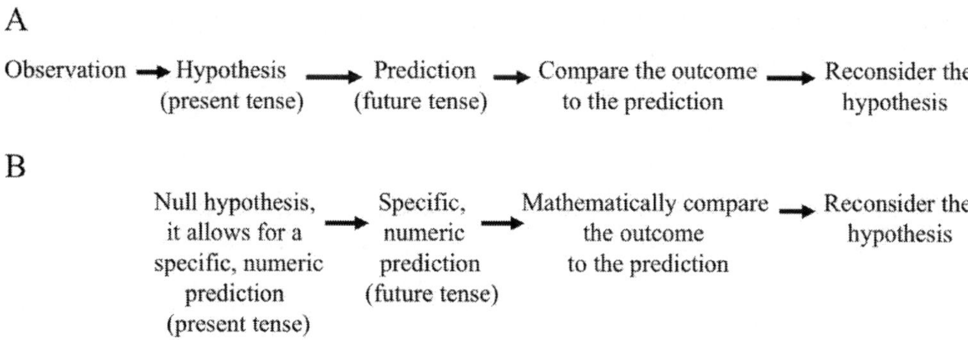

Figure 2.2 A. The hypothetico-deductive method. B. Null hypothesis testing.

We randomly choose ten CF-1 mice from our colony and determine their forage indexes. Those ten forage indexes represent the outcome of the study. We compare the outcome to the prediction. With that comparison in mind, we reconsider the hypothesis.

Null hypothesis testing is nothing more than a mathematical means of applying the hypothetico-deductive method. **It tells us how often sampling error alone will generate an outcome that differs from a prediction by a certain amount, or more**. If the answer is *rarely*, we generalize our finding to the population level. The finding is not due to sampling error. The finding is meaningful. We could be wrong, but that point is addressed in Section 2.6.

To apply a mathematical approach, it is necessary that the prediction be specific and numeric—and not a guess (Figure 2.2B). The prediction that the sample mean will be greater than zero does us no good. We have no idea how much greater it will be. A specific, numeric prediction would be that the sample mean will be zero, actually 0.000000000...; that is, to an infinite number of decimal places. Such a prediction requires a different hypothesis, in this case that CF-1 mice in our colony have no preference, on average, between Brand X and Brand W. In other words, our hypothesis is that the population mean is zero. This new hypothesis is a **null hypothesis** (or simply *null*), which is any hypothesis that allows for a specific, numeric prediction that is not a guess. Fisher stated that the null must be "exact" and "free from vagueness and ambiguity".[1] Otherwise, it is not possible to calculate a probability, which is where we are heading with this.

In some cases, the prediction is built into the mathematics of the null hypothesis test, so that the person conducting the test is unaware of the prediction, but **there is always a prediction** in null hypothesis testing.

[1] R. Fisher. 1971. The Design of Experiments. Hafner Publishing Company. p. 16.

2.3 What the Null Hypotheses in Our Statistics Books Really Mean

Our null hypothesis is that the CF-1 mice in our colony have no preference, on average, for either Brand X or Brand W (Chapter 1). If this were a "proper statistics book", how would that be stated? Probably as $\mu = 0$, with μ being the population mean. Thankfully, this is not a "proper statistics book", but we still need to consult such texts from time to time. Here are some null hypotheses in the language of statisticians, explained.

The samples were drawn from the same population.

There are two or more samples being compared to each other. If there is no effect of the independent variable, then the prediction that follows is that the sample means[1] will be exactly the same. Each sample mean would be an estimate of the same population mean. This null might translate as *this has no effect on that* or *all these things have the same effect on that.*

The population means are the same.

This means the same thing as the samples were drawn from the same population.

The samples were drawn from a population with a 1:1 ratio of this to that.

There are equal number of *this* and *that*. The prediction that follows is that there will be equal numbers of *this* and *that* in the sample.

The samples were drawn from a population with a 9:3:3:1 ratio this to this to this to that.

The example stems from Mendel's genetics experiments. If two genes are on separate chromosomes, the prediction that follows is that the outcome will take the form of a 9:3:3:1 ratio. This null translates as *these two genes are on separate chromosomes.*

2.4 The Meaning of *P*

The outcome of null hypothesis tests is *P,* which tells how often sampling error alone would generate an outcome that differs from a prediction by a certain amount, or more. In the case of mouse preference (Chapter 1), the outcome of 0.19 is different from zero.

---

[1] Or medians, or slopes, whatever.

If sampling error alone would rarely create an outcome that differed by 0.19 or more from zero, then the finding can be generalized to the population level. The finding is not due to sampling error. The finding is meaningful. The CF-1 mice in the colony prefer Brand X. We could be wrong, but that is explained in Section 2.6.

2.5 We Do Not Stop at Rejecting the Null. We Go One Step Further

It is usually stated that the goal of null hypothesis testing is to determine if the null can be rejected as false. If the null is *these two genes are on different chromosomes*, and we exclude chance, we reject the null, decide that they share a chromosome, and we are done. More often, when we reject a null, we accept a **directional difference** as meaningful. If sample mean A is greater than sample mean B, and we exclude chance, we can generalize to the population level, and decide that mean A is greater than mean B in general, and not the other way around. By accepting this direction, we go one step further than rejecting the null hypothesis.

Accepting a directional difference as meaningful does not require one-tailed testing, a topic covered at length in any "proper statistics book". Briefly, with one-tailed testing, we decide ahead of time that we expect a difference in a certain direction, we ignore differences in the opposite direction, and this affects the *P*-value favorably. Remarkably, Kaiser[1] successfully published his opinion we cannot accept a direction as meaningful without performing a one-tailed test. Bakan ridiculed Kaiser's idea as follows:

> *One really needs to strike oneself in the head! If Sample Mean A is greater than Sample Mean B, and there is reason to reject the null hypothesis, in what other direction can it reasonably be? What kind of logic is it that leads one to believe that it could be otherwise than that Population Mean A is greater than Population Mean B? We do not know whether Kaiser intended his paper as a reductio ad absurdum, but it certainly turned out that way.*[2]

See also Harris.[3]

2.5.1 In Many Cases, the Null Cannot Be Correct, But It Is Important to Test It Anyway

The belief that a null hypothesis can never be correct goes back at least to 1967.[4] Though that belief is not literally true, there are many cases in which a null cannot possibly be

[1] H. Kaiser. 1960. Directional statistical decision. Psychological Review, 67: 160–167.

[2] D. Bakan. 1966. Psychological Bulletin, 66: 423–437.

[3] R. Harris. 1997. Psychological Science, 8(1): 8–11

[4] P. Meehl. 1967. Philosophy of Science, 34: 151–159.

correct.[1] Because the prediction must be infinitely precise, **a null must also be infinitely precise.**[2] Consider the null *there are equal numbers of male and female squirrels on the campus of Lafayette College, in Pennsylvania*. That null could be true, but suppose one squirrel dies. Now it must be false. So, when populations are small and real, a null can be correct, but only trivially, and temporarily. How about experimental studies, in which populations are hypothetical? If the universe behaves deterministically, then everything must affect everything else, if only to a tiny degree. A null that states *this has no effect on that* must be wrong. If the universe does not behave deterministically, then *this has no effect on that* may be correct. Thankfully, the nature of the universe is irrelevant. Most often we ask if we can exclude sampling error and decide that a directional difference is meaningful (Section 2.5). This point was well stated by "the Picasso of statistics",[3] the legendary John Tukey. Here he assumes a deterministic universe and describes a comparison of two treatments, A and B.

> *All we know about the world tells us that the effects of A and B are always different—in some decimal place—for any A and B. Thus asking "Are the effects different?" is foolish. What we should be answering first is "Can we tell the direction in which the effects of A differ from the effects of B?" In other words, can we be confident about the direction from A to B? Is it "up", "down", or "uncertain"?*[4]

It does not matter if it is not possible for a null to be correct. A null hypothesis is only a tool.

2.6 How to Use *P*

P is a probability, so it ranges from zero to one. The lower the value of *P*, the less plausible it is that sampling error alone created a directional difference. We need a criterion value of *P* such that, should we calculate a *P*-value less than or equal to that criterion, we accept a directional difference as meaningful. That criterion is symbolized

[1] There is a large literature on this topic. For an overview, see J. Cohen. 1994. American Psychologist, 49(12): 997—1003.

[2] There is a large literature on this topic. e.g. D. Rindskopf. 1997. Chapter 12 in L. Harlow, S. Mulaik, and J. Steiger, eds. What if There Were no Significance Tests? Psychology Press.

[3] D. Salsberg. 2002. The Lady Tasting Tea. Henry Holt and Company, LLC. p. 230.

[4] J. Tukey. 1991. The philosophy of multiple comparisons. Statistical Science, 6: 100–116.

by α,[1] which is most often set to 0.05.[2,3] What does it mean to set α to some level, like 0.05? If we set α to 0.05, it means that, over a lifetime of excluding sampling error as the sole cause of a directional difference, sampling error will be the sole cause of that directional difference 5% of the time. In other words, 5% of the time we say *this is a meaningful result* we will be wrong. Making that mistake is referred to as a **type I error**, which is usually described as rejecting a true null hypothesis. Since it is generally irrelevant whether or not the null is correct (Section 2.5), we can think of a type I error as accepting a directional difference as meaningful when it is not. The criterion α is the probability of a type I error over a lifetime of testing null hypotheses.

It is important to recognize that a 5% type I error rate assumes that everything about a research study is perfect. In practice, subtle procedural imperfections can introduce lurking variables, which bias the results. So, even if α is set to 0.05, the actual error rate over a lifetime of testing nulls is some higher, unknown number. An unexpected finding that $P \leq \alpha$ is especially worrisome. Methods should be scrutinized for flaws. If none are found, 0.05 is still 0.05. Inevitably, we will make the mistake of declaring meaningless results meaningful. Abelson stated this well.[4]

> *Let us stop viewing statistical analysis as a sanctification process. We are awash in a sea of uncertainty, caused by a flood tide of sampling and measurement errors, and there are no objective procedures that avoid human judgment and guarantee correct interpretations of results.*

When we know the null cannot be true, or it can only be true in transient, unimportant ways (Section 2.5.1), we should use *P* as follows:

If $P \leq \alpha$, we

- conclude that the direction of a difference observed in our results is the same as the direction of a difference at the population level, e.g., the CF-1 mice in our colony prefer Brand X over Brand W, and not the other way around,

---

[1] Fisher used *sig*, but I prefer Neyman and Pearson's α for the reasons explained in Section 3.1

[2] Why 0.05? Fisher (see the Preface) strongly advocated the criterion of 0.05. Later, he changed his mind, and decided that α should be decided on a case-by-case basis. It was too late. The value of 0.05 had become institutionalized.

[3] I follow the Council of Science Editors' recommendation and put a zero before the decimal place. Some omit the zero, because a probability cannot be greater than one.

[4] R. Abelson. 1997. Psychological Science, 8: 12-15.

- exclude sampling error as the sole reason for the direction of a difference in our results,
- consider our results meaningful,
- and keep in mind that, if $\alpha = 0.05$, we will be incorrect in doing these things 5% of the time, and that is only if our studies are carried out to perfection. This is not a "sanctification process."[1]

If $P > \alpha$, but it is close to α, we form no conclusion, and consider that

- the direction of a difference observed in our results *may* be the same as the direction of a difference at the population level,
- sampling error may not be the sole reason for the direction of a difference in our results,
- the results *may* be meaningful,
- we might want to increase the sample size (Section 2.7), or repeat the study with a different strategy (Chapter 15), because either might lead to an outcome in which $P \leq \alpha$.

If $P >> \alpha$, we

- cannot conclude that the direction of a difference observed in our results is the same direction of a difference at the population level,
- cannot exclude sampling error as the sole cause of the direction of a difference observed in our results,
- and consider that any directional difference at the population level may be so small it is not worth trying to identify its direction by increasing the sample size or repeating the study differently.

Suppose the null is *these two genes are on different chromosomes*. There is no directional difference. We can use P in a fashion similar to above, but only to decide whether we should reject the null or draw no conclusion, i.e., fail to reject. An outcome of $P > \alpha$ does not mean we should accept the null hypothesis (Section 3.3).

2.7 If an Outcome is Meaningful, How Do We Ensure We Identify It as Meaningful?

Suppose we have a directional difference, and it is meaningful; it is not the result of sampling error. We need to identify it as such. We need an outcome of $P \leq \alpha$. Otherwise,

[1] R. Abelson. 1997. Psychological Science, 8: 12-15.

we have failed. Such a failure might be called a **type II error**. Traditionally, however, a type II error is said to occur when we *accept* a null that is incorrect. We can avoid such an error by never accepting nulls and failing to reject instead. We might also think of a type II error as failing to reject a null that is incorrect. But what if we know ahead of time that the null must be false? We could redefine *type II error* as failing to recognize a directional difference as meaningful. We do not need a third definition of type II error. However we define this error, we need to avoid it.

A test's ability to generate an outcome of $P \leq \alpha$ when a directional difference is meaningful is referred to as the test's **power**, though power is traditionally defined in terms of rejecting a null that is incorrect (avoiding a type II error). **Many aspects of research design can be used to maximize power**. One way to maximize power is to use the largest samples sizes possible. Also, it is best to have equal sample sizes, as many tests are most powerful when sample sizes are equal (for an exception, see Section 9.3.1). Another strategy, appropriate for experiments, is to manipulate the independent variable to create the largest difference between a prediction and an outcome. Finally, for numerical data, there is variation. The more scatter in the data, the less confident we can be in them, the lower the power, and the higher the *P*-value. Various strategies can be used to reduce scatter and increase power. Those strategies include the addition of another variable in an ANOVA (Section 10.7), adding a covariate to create an ANCOVA (Chapter 11), and using a repeated measures design (Chapter 13). The difference between success and failure often comes down to the intelligent application of those three statistical techniques.

2.8 Never Do This

A common mistake is to inspect data, see some directional difference, and then use those same data to test a hypothesis that is inspired by that directional difference. Doing so is referred to as overfitting,[1] and it is invalid to overfit. Imagine that we see a cloud that looks like a horse, we form the hypothesis that all clouds look like horses, and we test that hypothesis by looking at the same cloud. Testing a hypothesis with the same data that inspired that hypothesis is just as bad. Consider the following ten random samples from the same population (Figure 2.2). If *C* and *F* represent different treatments, we might decide that those treatments have different effects. This is a new research

[1] G. Gigerenzer. 2004. The Journal of Socio-Economics, 33: 587–606.

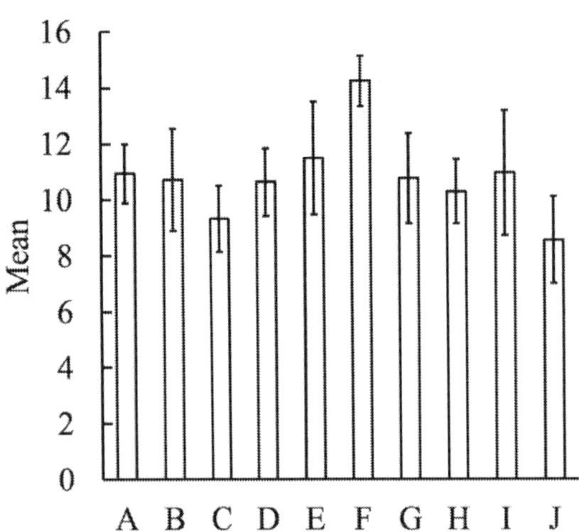

Figure 2.2 Ten random samples of a population with a mean of 10 and a standard deviation of five. Bars show standard error of the mean.

hypothesis, inspired by our results. If we use a *t*-test to compare results from groups *C* and *F*, we find that $P = 0.003$. Naturally, the results that inspired the hypothesis that *C* and *F* have different effects also support that hypothesis. But the samples were all drawn from the same population, so sampling error alone created the difference between groups *C* and *F*.

Results like those in Figure 2.2 need to be analyzed in an unbiased fashion; all samples should be compared to each other.[1] This can be accomplished with analysis of variance. With the data in Figure 2.2, the result would be that $P = 0.498$.

If we can only generate one set of data, and we plan to use it to both generate a hypothesis and test that hypothesis, what do we do? We use randomization without replacement (Section 1.3) to assign each datum to one of two groups. We use one group to formulate a research hypothesis, then use the other group to test it with a null hypothesis. The sample size is cut in half, and that is the downside of this strategy.

2.9 Null Hypothesis Testing Explained as Never Before

In null hypothesis testing, it is not necessary to make observations and then form a research hypothesis (Figure 2.2). We may have no idea which brand our CF-1 mice prefer (Chapter 1). Perhaps Brand X was developed with better nutrition in mind, rather than better flavor. Suppose we obtain a sample mean of 0.19, find that $P \leq \alpha$, and conclude that the CF-1 mice in our colony prefer Brand X over Brand W, and not the other way around. This is a new hypothesis, inspired by our results. What null

[1] Some authors argue against making all possible pair-wise comparisons, e.g. L. Wilkinson and the Task Force on Statistical Inference. 1999. American Psychologist, 54(8): 594–604.

hypothesis testing tells us is that, over a lifetime of formulating such data-based hypotheses, we will only be wrong 5% of the time.[1] We have no research hypothesis, we collect data, observe a difference with a certain direction, and that directional difference is self-proven, provided that $P \leq 0.05$, and provided that we are willing to take that 5% risk. Do we not need to test such data-based hypotheses with new data (Section 2.8)? No. This is not the same as deciding to compare groups *C* and *F* and ignoring the others. Does this not require one-tailed testing? No (Section 2.8).

[1] Of course, this assumes that we set α to 0.05.

CHAPTER 3
COMMON MISCONCEPTIONS CONCERNING NULL HYPOTHESIS TESTING

3.1 Statistical Significance Means a Difference Is Large Enough to Be Important—Wrong!

Historically, results have been described as *significant* whenever $P \leq \alpha$, but significant only in the sense that the results signify something.[1] A finding of $P \leq \alpha$ does not in itself mean that results are important. It is commonly found that, with very large sample sizes, $P \leq \alpha$ for trivial differences. For example, suppose we poll 15,000 voters in Dade County, Florida, and another 15,000 voters in Palm Beach County, and ask their party affiliations. If we find that 66% of them are Democrats in Dade County, and 64% are Democrats in Palm Beach County, the result is that $P \leq 0.05$.[2] But how important is a difference of 2%? It is not important, but it would qualify as statistically significant.

There is probably a suite of subtle reasons why a few more Democrats might live in Dade County than Palm Beach County. Democrats may be drawn to Dade a little more than Palm Beach, or maybe growing up in Dade makes people a little more prone to siding with the Democrats. If we consider small populations, though, things get even worse when it comes to testing nulls. Consider the null hypothesis that there are equal numbers of male and female squirrels on the campus of Lafayette College, in Pennsylvania. Most likely, that null is false, and if we examine enough squirrels, we would declare the outcome significant. But there may be no reason why there are more of one sex than the other. Nulls can be incorrect because of chance alone. **Testing nulls cannot tell us whether a null is wrong due to chance alone or due to some systematic process**.

The phrase **significantly different** is especially **misleading** because it implies that we test nulls to decide if differences are large enough to be important—important

---

[1] D. Salsberg. 2002. The Lady Tasting Tea. Henry Holt and Company, LLC. p. 98.

[2] $X^2 = 4.62$.

because of not being created by chance alone, or important in some broader sense. Traditionally, we test nulls to reject or fail to reject the null hypothesis, and **a null hypothesis cannot be wrong to different degrees**. We do not test nulls to see if they are wrong enough for a finding to be important. We test them to see if we can exclude sampling error as the cause of some difference.

A better word than *significant* is *meaningful*. *Meaningful* only implies that results mean something, not that they are important or large in magnitude.[1]

3.2 *P* is the Probability of a Type I Error—Wrong!

Historically, rejecting a correct null hypothesis has been referred to as a **type I error**. In most cases, we can think of a type I error is accepting the direction of a difference as meaningful when it is not. Regardless, we either make a type I error or we do not. Its probability in any one circumstance is either one, if we make the error, or zero, if we do not. The probability of making a type I error is α, not P, and it is over a lifetime of testing null hypotheses, not in any one instance.

Similarly, ***P* is not the probability the null is correct**. The null is either right or wrong. The direction of a difference is either meaningful or it is not. Again, we have probabilities of one and zero, but nothing in between.

Gigerenzer[2] traces these misconceptions concerning P back to Guilford's *Fundamental Statistics in Psychology and Education*, which was published in 1942. Gigerenzer goes on to describe how these misconceptions were passed down through generations of authors.

3.3 If P>α, We Should Accept the Null—Wrong!

Fishere was adamant that the null should never be accepted,[3] but accepting the null was part of Neyman and Pearson's theorem. Thanks to Neyman and Pearson, many authors and teachers have said that the null should be accepted if $P > a$. But it is the persons who contribute to the primary literature who are the real authorities. Here is what they say.

> *The belief that "if the null hypothesis is not rejected, then it is to be accepted . . . [is] the most devastating of all to the research enterprise."*[4]

---

[1] "We in the behavioral sciences should 'give' this word [significance] back to the general public." R. Kline. 2004. Beyond Significance Testing: Reforming Data Analysis Methods in Behavioral Research. American Psychological Association, 325 pp. See p. 87.

[2] G. Gigerenzer. 2004. The Journal of Socio-Economics, 33: 587–606.

[3] Ibid. pp. 107,108.

[4] F. Schmidt. 1996. Psychological Methods, 1: 115–129.

> *"Failing to reject the null hypothesis essentially provides almost no information about the state of the world. It simply means that given the evidence at hand one cannot make an assertion about some relationship: all you can conclude is that you can't conclude that the null is false."*[1]

> *"The worst, i.e., most dangerous feature of 'accepting the null hypothesis' is the giving up of explicit uncertainty: the attempt to paint with only the black of perfect equality and the white of demonstrated direction of inequality. Mathematics can sometimes be put in such black-and-white terms, but our knowledge or belief about the external world never can."*[2]

Tacit acceptance of the null is rampant. Anytime we find that $P > \alpha$ and conclude that a treatment has no effect, or that multiple treatments have the same effect, we are tacitly accepting the null and reaching an unjustified conclusion. Doing so is very common, thus, "the most devastating practice of all to the research enterprise."

3.4 Based on *P* We Should Either Reject or Fail to Reject the Null—It Is Not That Simple

Null hypothesis tests should not be used to make mindless, binary decisions. We should use *P*-values to think more about our data, not less. In some fields, the use of *P*-values to make mindless decisions progressed to where results were often not shown in figures or tables, but only as $P \leq \alpha$, *reject* or $P > \alpha$, *fail to reject* (or worse, *accept*). Also, in those fields, the phrase is typically *null hypothesis significance testing*, initialized to NHST. In the minds of many, NHST implies making mindless decisions to accept or reject, while not thinking about the data. It is best that we not use the initialism NHST, or write the phrase out in full, unless we want to be painted as part of the problem.

3.5 *Power* Can Be Used to Justify Accepting the Null Hypothesis—Wrong!

Power is one of Neyman and Pearson's embellishments to Fisher's method. Power refers to a test's ability to detect as meaningful a directional difference that is not caused by sampling error (Section 2.7). It is the ability of a test to flag a meaningful result as meaningful. With power calculated, it is possible to determine the sample size necessary to achieve the outcome of $P \leq \alpha$, given that the null is incorrect to a certain, minimum

---

[1] Gill. Political Research Quarterly, 52: 647–674, quoting Cohen, Journal of Abnormal and Social Psychology, 65: 145–153.

[2] J. Tukey. 1991. Statistical Science, 6: 100–116.

degree. For example, if the null is that the mean foraging index of all the CF-1 mice in our colony is zero (Chapter 1), it is possible to use power to determine the sample size needed to achieve an outcome of $P \leq \alpha$ if the actual mean forage index differs from zero by 0.2 or more. It is a mistake to think that, if we have that sample size, and $P > \alpha$, we can accept the null hypothesis. In this example, such a finding would only indicate that the null is not incorrect by 0.2 or more.

3.6 If $P \leq \alpha$, We Should Accept the Alternative Hypothesis—It Is Not That Simple

We are often taught that the alternative, or alternate, hypothesis comprises everything other than the null. So defined, the alternative hypothesis is useless. Here is why. The alternative hypothesis was created by Neyman and Pearson, because they needed it to determine power (Section 2.7). Continuing from Section 3.5, the alternative hypothesis would be that the CF-1 mice in our colony have an average preference 0.2 or greater. If that alternative hypothesis were true, we could use power to calculate the sample size necessary to reject the null and accept the alternative, provided that the null is wrong by 0.2 or greater. Neyman and Pearson were inconsistent in how they described the alternative hypothesis—from highly specific to *everything other than the null*—but the minimum difference between the null and the alternative must be specified[1] (as well as the shape of both distributions, see Figure 4 in Perezgonzalez[1]). The alternative hypothesis *the mice have a preference* would be useless. With no minimum difference specified, there would be no power, and the sample size necessary would be infinite.

3.7 The Null Hypothesis Is a Statement of No Difference—Not Always

A null hypothesis is any hypothesis that allows for a specific, numeric prediction that is not a guess. A null hypothesis is called a null hypothesis because it may be nullified.[2] A statement of no difference is best described as a **nil hypothesis**.[2] Many null hypotheses are also nil hypotheses. For example, *there is no difference in the effectiveness of these two chemotherapy drugs.* That hypothesis is a statement of no difference, and it also allows for a specific, numeric prediction such as *five-year survival rates will the same regardless of whether patients get this chemotherapy drug or that one.* But a hypothesis may allow for a specific, numeric prediction without being a statement of no difference. An example would be *these two genes are on different*

[1] J. Perezgonzalez. 2015. Frontiers in Psychology, 6: article 223.

[2] J. Cohen. 1994. American Psychologist, 49: 997–1003.

chromosomes. That null allows for the prediction that the outcome will take the form of a 9:3:3:1 ratio. So, all nils are nulls but not all nulls are nils.

3.8 The Null Hypothesis Is That There Will Be No Significant Difference Between the Expected and Observed Values—Very, Very Wrong!

Consider the logic of testing null hypotheses (Section 2.2) and how the "null" above turns that logic inside out. In chi-squared tests, the expected values (or frequencies) make up the prediction that is based on the null hypothesis being true. If the prediction follows from the null, how can that prediction be part of the null? Also, suppose the outcome is $P \leq \alpha$. What would it mean? Since this "null" never addresses the population level, there is no way to generalize the finding (Chapter 2). This sort of "null" never appears in "proper statistics books", but it crops up in the context of biology teaching laboratories. I can only imagine this is because students can spit it back without thinking, anytime they perform a chi-squared test.

3.9 A Null Hypothesis Should Not Be a Negative Statement—Wrong!

A null hypothesis is any hypothesis that allows for a specific, numerical prediction (Section 2.2). I am told that the objection to negative nulls is that we should not reject a negative statement. Fisher did not share this concern. Consider his example of the woman who claimed she could tell by taste whether milk or tea was poured into a cup first. He stated the null as "the judgements given are in no way influenced by the order in which the ingredients have been added"[1]—a negative statement.

---

[1] R. Fisher. 1971. The Design of Experiments. Hafner Publishing Company. pp. 15,16.

CHAPTER 4
CRITICISMS AND WISE USE

4.1 The Criticisms

There has been a long and fierce debate over null hypothesis testing.[1] Criticisms include the following.

- A null hypothesis can never be correct (2.5.1). Why test it? If we fail to reject, it only means that there is insufficient power (Section 2.7).
- Null hypothesis tests are used to make mindless, binary decisions.
- Focus on the results of null hypothesis tests comes at the expense of other important aspects of the data, such as effect size.
- Results of null hypothesis tests are often misinterpreted. Many think that, if $P \leq \alpha$, a difference is large enough to be important (Section 3.1). Many accept the null hypothesis if $P > \alpha$, when they should be forming no conclusion (Section 3.3).
- Other means of drawing conclusions, such as through Bayesian inference, are superior.

4.2 Wise Use Is the Answer

Many of those criticisms can be addressed through *wise use* of null hypothesis testing.

- It may be that the null is almost always incorrect, but the null is only a tool. If the null cannot be correct, the reason to test it is to decide if we should accept the direction of a difference as meaningful. It is also important to remember that some null hypotheses *can* be correct (Section 2.5).
- Null hypothesis testing should not be used to make mindless, binary decisions. If $P > \alpha$, but it is close to α, it is important to consider that the direction of a difference may be meaningful, but perhaps there was insufficient power to detect it (Section

---

[1] Much of the debate is encapsulated by L. Harlow, S. Mulaik, and J. Steiger, eds. 1997. What If There Were no Significance Tests? Psychology Press.

2.7). This is one reason it is important to **always provide precise**[1] ***P*-values**. An outcome of $P = 0.056$ is very different from an outcome of $P = 0.56$, since the former is so close to α. Precise *P*-values allow readers to draw their own conclusions.

- Results should be presented in the form of graphs and tables; with means, medians, and such; along with confidence intervals (Chapter 14) or measures of scatter, such as quartiles; so readers can draw their own conclusions. Section 14.5, in particular, illustrates an excellent way to display data and how they might be interpreted.
- The null should not be accepted unless something like Frick's[2] guidelines (Chapter 5) are followed, and $P > \alpha$ is not one of them. We should not conclude that one thing has no effect on another, or that two things have the same effect, as doing so is tacit acceptance of the null.
- **Results should not be described in terms of *significance***, as the word is too often misinterpreted (Section 3.1).[3] Instead we can state that we can or cannot exclude sampling error, or we can or cannot attribute an effect to an independent variable. If it is important to use one word, results should be described as *meaningful* rather than *significant*.
- We should not be so dependent on *P*-values that we are afraid to use our own intuition and intelligence to draw conclusions. We should not be so dependent that we ignore questions that cannot be answered by testing a null.
- We should keep in mind that there are many strategies for inference, and null hypothesis testing is only one of them. Null hypothesis tests help us understand single sets of data.
- It is often important to focus on effect size (Chapter 6), the magnitude of an effect that one variable has on another. Effect size is especially important when studying natural variation. When effect size is important, it should not be neglected in favor of testing nulls.
- Nulls should only be tested when data are ambiguous. Nulls should only be tested to ask important questions. We should not evaluate every possible research finding, no matter how trivial, by testing a null. Doing so creates *P*-clutter,[4] a results section that is full of unimportant *P*-values. Such a results section lacks focus and is a burden to read.

---

[1] If we are using a table to find *P*, we should provide a narrow range, e.g., $0.10 > P > 0.05$.

[2] R. Frick. 1995. Memory & Cognition, 23: 132–138.

[3] *Significant* only means that the results signify something. D. Salsberg. 2002. The Lady Tasting Tea. Henry Holt and Company, LLC. p. 98.

[4] R. Ableson. 2009. Chapter 5 in Harlow, S. Mulaik, and J. Steiger, eds. 2009. What if There Were no Significance Tests? Psychology Press.

- We should not state that *these groups do not differ*, when the sample means are almost certainly different.

4.3 Wise Use, Quiz Yourself

Imagine that we have a drug that we hope will improve memory. We randomly assign CF-1 mice from our colony to receive that drug. The day before the drug is administered, each mouse is given a memory task, and we record its score. The next day, the mice are tested again, but when the drug is in their systems. The day after that, the mice are tested one more time, after the drug is metabolized. Assume we perform the correct analysis[1]. How should we interpret the results (Figure 4.1)?

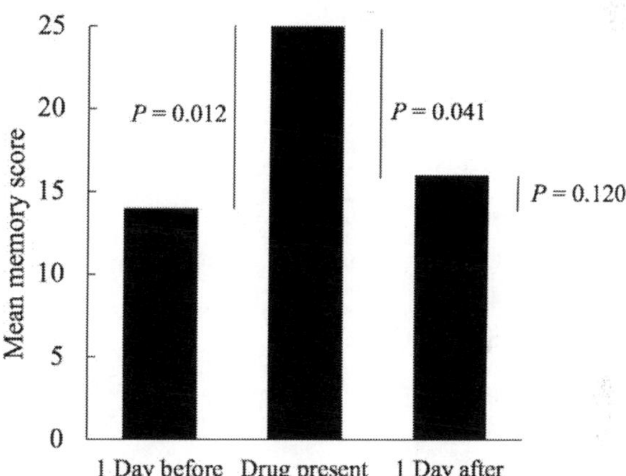

Figure 4.1 One set of mouse memory scores.

The following are poor interpretations. Try to figure out why, then read the explanations that follow.

Results of the drug were reversible because, when we compare day 1 to day 3, there is no difference.

This is wrong for several reasons. There is a difference in sample means, so it is wrong to say there is no difference. If we add mention of *significance* to clarify, the reader is likely to equate it with importance, when importance is better equated with effect size. Finally, it is tacit acceptance of the null hypothesis to say that, since $P > \alpha$, there is no difference in population means. In experiments like this, in which there is an attempt at

[1] A repeated measures ANOVA followed by post hoc tests with the appropriate error term for a repeated measures design

reversal of treatment (wash out, recovery, rescue, and so on), there is no reason to compare *before* with *after* by testing a null.[1]

Results of the drug were reversible because, when we compare day 2 to day 3, P = 0.041.

We cannot say they are *fully* reversable. Also, it is important to remember the 5% risk that we are wrong to exclude sampling error. In this case, *P* is close to 0.05, so it is especially important to keep that 5% in mind.

Suppose we get the results in Figure 4.2. The following is a poor interpretation. Try to figure out why, then read the explanations that follow.

The effect of the drug is long-lasting. There was no reversal a day after its administration.

Saying that there is no reversal, because $P = 0.120$, is tacitly accepting the null. A value of 0.120 suggests that power may have been too low (Section 2.7) for us to detect a reversal that really took place. We should also step away from *P*-values and use our intelligence and intuition. The purpose of a removal of treatment design is to provide

Figure 4.2 A second set of mouse memory scores.

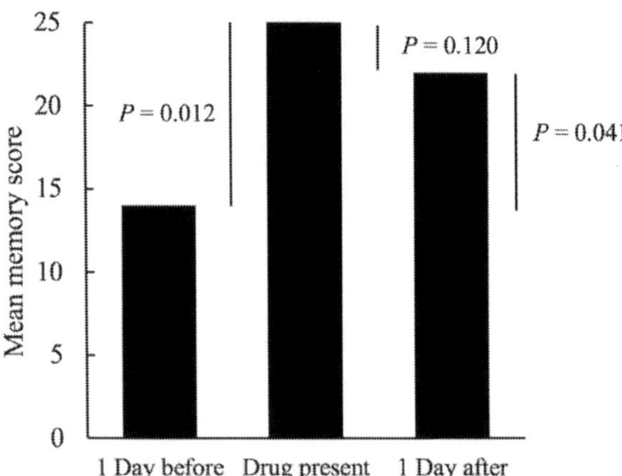

confidence that the initial change is brought about by the by the treatment, provided that the treatment can be reversed. Here, there is no convincing reversal. Perhaps something went wrong earlier, which suppressed the first set of scores. If so, there was no convincing recovery because the drug never had an effect in the first place.

[1] Unless we use an approach similar to Frick's, which is described in Chapter 5. R. Frick. 1995. Memory & Cognition, 23: 132–138.

CHAPTER 5
WHAT IF IS IMPORTANT TO ACCEPT THE NULL?

If it is important to accept the null, Frick[1] provides the following set of rules.

- "While a *P*-value of less than 0.20 seems too low [to accept the null], a *P*-value of greater than 0.50 seems large enough. The range of 0.20 to 0.50 currently seems to be ambiguous . . ." A finding that that $P > \alpha$ does not justify accepting the null.
- "That the results are consistent with the null hypothesis . . ."
- "That the experiment was a good effort to find an effect . . ." An example would be the creation of large samples.
- "That the null hypothesis is possible . . ." If a population is small and real, a null may be correct (Section 2.5.1). Suppose the population is hypothetical, as would be the case in an experiment. If we reject determinism, then one thing may have no effect on another, and a null could be correct. Suppose the null is *these two genes are on separate chromosomes*. That null could also be correct.

[1] R. Frick. 1995. Memory & Cognition, 23: 132–138.

CHAPTER 6
EFFECT SIZE: WHAT IS IT AND WHEN IS IT IMPORTANT?

Effect size is the magnitude of the effect that one variable has on another. Estimates of effect size may take simple forms, such as the difference between a prediction and an outcome. Effect size may also be estimated by more sophisticated quantities, such as the amount of variation in a dependent variable that can be attributed to an independent variable. Effect size can be of extreme importance and, while effect size influences P (the greater the difference between a prediction and an outcome, the lower the P-value; Section 7.7.1), P is influenced by other things as well. Therefore, P does not provide an estimate of effect size. We cannot say that the lower the P-value the more important the result (Section 2.7 and 7.11).

Effect size is most important when studying natural variation, e.g., *this affects that to this degree*. In contrast, in experiments, we often manipulate independent variables to extremes, to create the greatest differences possible in our outcomes. Since effect size influences P, we do so in the hope of finding that $P \leq \alpha$. There is nothing wrong with doing so. The aim is to show proof of principle, i.e., this *can* affect that. But it is important to remember that we have **intentionally created a large effect size**. On the other hand, if we are studying natural variation, e.g., determining how day length affects when a species' flowers first open, then effect size is important. It would let us know how important day length is in comparison to, for example, temperature.

When effect size is important, it is necessary to **report it, even if $P > \alpha$**. Why report the size of an effect when that effect could have been created by sampling error and not be an effect at all? Samples only allow for the estimation of effect size. Half of the time effect size will be overestimated, and half of the it will be underestimated. Because large effect sizes favor small P-values, if we only report effect size when $P \leq \alpha$, effect size will be overestimated more often than underestimated. This phenomenon is termed the *winner's curse.*[1] Effect sizes, overall, would be inflated.

We assess effect size anytime we compare an outcome with a prediction. We may also estimate effect size with R^2, which indicates the degree to which an independent variable influences a dependent variable. When it comes to other measures of effect size, we are left with a literature that is largely ignored by "proper statistics books". Good places to start are Cohen[2] and Fritz et al.[3]

---

[1] Z. Zöllner and J. Pritchard. 2007. Overcoming winner's curse: estimating penetrance parameters from case control data. The American Journal of Human Genetics, 80: 605–615.

[2] J. Cohen. 1992. Psychological Bulletin, 112(1): 155–159.

[3] C. Fritz, P. Morris, and J. Richler. 2011. Journal of Experimental Psychology: General, 141(1): 2–18.

CHAPTER 7
SIMPLE PRINCIPLES BEHIND THE MATHEMATICS AND SOME ESSENTIAL CONCEPTS

7.1 Why Different Types of Data Require Different Types of Tests

7.1.1 Simple Principles Behind the Mathematics

The greater the difference between a prediction and an outcome, and the greater the sample size, the more confident we can be that our outcome is meaningful. Thus, the result of null hypothesis tests is influenced by sample size and the difference between the prediction and the outcome. Large sample sizes, and large difference between the prediction and the outcome, favor low *P*-values. On the other hand, the greater the variability (scatter) within our results, the less confident we can be in drawing conclusions from them—but not all types of data exhibit variation. So, some null hypothesis tests are designed to favor low *P*-values when there is little scatter in the data, while others are designed for data that exhibit no scatter at all. To choose an appropriate test, it is necessary to first understand why some data exhibit scatter and some do not. Data may be broadly categorized as nominal or numerical.[1]

7.1.2 Numerical Data Exhibit Variation

Numerical data often consist of measurements (continuous numerical data). If we randomly collect ten clams from below a sewage treatment plant, and determine their dry weights, those dry weights would make up a sample of numerical data. Numerical data may also consist of a set of counts (discrete numerical data). For example, we may be interested in the ability of the single-celled organism, *Tetrahymena*, to ingest latex beads. We immerse the cells in a solution that contains a certain concentration of beads, wait a period of time, randomly select 20 cells, and count the number of beads in each.

---

[1] Any "proper statistics book" would provide more types of data, such as those on an ordinal scale.

The resulting sample would consist of discrete numerical data, since you cannot have a fraction of a bead.

Both in the case of the clams and in the case of *Tetrahymena*, we could calculate an average, perhaps it is 4.7 g and 5.8 latex beads. And it is not like every clam we collected weighed 4.7 g. It is not even possible to consume 5.8 latex beads. The data would be dispersed around those averages. Our results would exhibit variation—scatter. With numerical data, the correct test is often a type of *t* test or analysis of variance (ANOVA),[1] because those tests generate outcomes that are influenced by scatter in the data, along with sample size, and the difference between the prediction and the outcome.

7.1.3 Nominal Data Do Not Exhibit Variation

Nominal (or **enumeration**) **data** are counts of observations within categories. We may wonder whether party affiliation differs between Dade and Palm Beach Counties in Florida. We poll about 100 voters at random and ask their party affiliations. The result would be nominal data, counts of observations within categories (Table 7.1).

Table 7.1 Party affiliations in two Florida counties

County	Democrats	Republicans	Other
Dade County	65	14	21
Palm Beach County	75	2	18

With nominal data, there is no way to calculate an average. The concept of variation does not apply. We often use types of chi-squared tests for nominal data. The outcome of chi-squared tests is only influenced by the sample size and the difference between the prediction and the outcome.

7.1.4 How to Tell the Difference Between Nominal and Numerical Data

At this point you should be confused. Counts may be nominal data, or they may be numerical. Nominal data take the form of numbers, but they are not considered numerical. The solution is to ask, *can I take an average?* If we are asking the party affiliation of persons in two Florida counties, we cannot. The data are nominal. If we are weighing clams that we collected below a sewage treatment plant, we might calculate an average of 4.7 g. The data are numerical.

[1] *t* tests are actually special cases of ANOVAs.

7.2 Simple Principles Behind the Analysis of Groups of Measurements and Discrete Numerical Data

7.2.1 Variance: A Statistic of Huge Importance

For numerical data, the test statistic, such as t or F, often combines three aspects of the data into one number: the difference between the prediction and the outcome, the amount of scatter in the data, and the sample size. To illustrate how tests accomplish this, we will analyze the data on mouse preference (Example 2.1) with a **single-sample *t* test**,[1] illustrated in Example 7.1.

Example 7.1 Forage indexes from Example 2.1 and a single sample t test.

	Forage index (x)	$(x-\bar{x})^2$		
	0.12	0.005		
	-0.09	0.078		
	0.14	0.003		
	-0.08	0.073		
	0.31	0.014		
	0.10	0.008		
	0.76	0.325		
	0.44	0.063		
	-0.03	0.048		
	0.12	0.005	n =	10
mean ($\bar{x}$) =	0.19		df =	9
	SS =	0.617	s^2 =	0.069

The symbol x and refers to the data in the sample. The sample mean is $\bar{x}$. The sample size is n, and degrees of freedom, $n-1$, is shown as *df.* **A statistic of great importance is variance** (s^2 or *error MS*), which indicates the amount of scatter within one or more samples. To obtain variance for a single sample, we start by subtracting the sample mean from each datum (or the other way around) and squaring the difference. Here, we start with $(0.019 - 0.012)^2 = 0.005$, $(0.019 - (-0.009))^2 = 0.078$, and so on; thus, the second column shows squared deviations from the mean. Summing those squared deviations gives us the sum of the squares, *SS*, literally the sum of the squared deviations from the mean. Dividing the sums of the squares by the number of degrees of freedom gives us **variance, the average squared deviation from the mean**. The importance of this

[1] The fact that Ivlev's forage ratio is subject to a floor effect is addressed in Section12.1.

awkward quantity cannot be overstated. Variance is used in many statistical tests as an indicator of scatter.

Variance is also termed *mean square error, mean square*, *the error term* or *error*. The word *error* does not mean anyone made a mistake. While measurement error contributes to variance, **error can mean natural variation** in what is being measured.[1] In biological data, error is almost entirely the result of natural variation.

If variance is the average squared deviation from the mean, why obtain it by dividing the sum of the squares by degrees of freedom? Why not use the sample size? Using degrees of freedom corrects for a problem caused by small sample sizes.[2] The smaller the sample, the greater the effect of subtracting one, and the greater the correction. Why square the deviations to make them positive, when it is just as easy to use their absolute values? The result of doing the latter would be *mean deviation*, which is hardly ever used in statistics. Why not take the square root of variance, so as to recover the original units of measurement? This is done. The square root of variance is **standard deviation**, the square root of the average squared deviation from the mean, an even more awkward quantity than variance. Standard deviation is also referred to as *root mean square*.

7.2.2 Incorporating Sample Size and the Difference Between the Prediction and the Outcome

The next step is to calculate standard error of the mean, *SE,* as follows.

$$SE = \sqrt{\frac{s^2}{n}}$$

$$SE = 0.083$$

Note that a large amount of scatter, and a small sample size, both favor large values of standard error. Large values of standard error suggest that we should not be confident in drawing conclusions from our data. The opposite is true if standard error is low.

The next step is to factor in the difference between the prediction and the outcome. This is accomplished by the numerator in the equation below, since μ represents the prediction based on the null hypothesis being correct, and this is nearly always zero. The result is t, a number that combines all three aspects of the data into one quantity.

---

[1] See D. Salsburg. 2001. The Lady Tasting Tea: How Statistics Revolutionized Science in the Twentieth Century. Henry Holt and Company, LLC, p. 16.

[2] The data will naturally be closer to the sample mean than the population mean, since the sample mean is calculated from those data. If sample size were used instead of degrees of freedom, variance would be underestimated.

$$t = \frac{|\bar{x} - \mu|}{SE}$$

$$t = 2.29$$

Recall that a low value of standard error suggests little scatter in the data, a large sample size, or both. A low value of standard error would tend to make *t* large. The value of *t* would also be large if there were a large difference between the prediction and the outcome. So, *t* is a combined representation of sample size, scatter, and the difference between the prediction and the outcome. A high value of *t* suggests that we should exclude sampling error as the sole cause for a directional difference, a low value the opposite. Fittingly, the higher the *t*-value, the lower the *P*-value.

Lastly, it is time to determine *P*. For that we need degrees of freedom again. As before, degrees of freedom functions as a correction factor for an issue caused by low sample size.[1] For a single-sample *t* test, *df* = *DF* (cf. Chapter 8). We locate our value of *DF* in the left-most column of Table 7.2, read across to the right to approximate the location of *t*, then read up to the top row for the *P*-value. Our *t*-value of 2.29 is in between 2.262 and 3.250, so *P* is in between 0.01 and 0.05.

Table 7.2. A probability table for *t*

	P (2-tailed)					
DF	0.50	0.20	0.10	0.05	0.01	0.001
7	0.711	1.415	1.895	2.365	3.499	5.408
8	0.706	1.397	1.860	2.306	3.355	5.041
9	0.703	1.383	1.833	2.262	3.250	4.781
10	0.700	1.372	1.812	2.228	3.169	4.587
11	0.697	1.363	1.796	2.201	3.106	4.437

7.3 Drawing Conclusions When We Knew All Along That the Null Must Be False

In this case, the outcome is that $0.05 > P > 0.01$. Since it is practically impossible that the mean forage index of all the CF-1 mice in our colony is zero to an infinite number of decimal places, it was clear from the start that the null must be false (Section 2.5).

[1] The *t* distribution is not normal, but it approaches being normal with larger and larger sample sizes. Having different rows in probability tables provides *P*-values that are corrected for the different shapes of the *t* distribution. The smaller the sample size, the greater the effect of subtracting one, and the greater the correction.

Because we generally set α to 0.05, and $P < 0.05$, we conclude that the CF-1 mice in our colony prefer Brand X over Brand W, and not the other way around. In setting α to 0.05, we would only be mistaken in drawing such a conclusion 5% of the time, over our lifetimes, provided that our studies are carried out flawlessly (Section 2.6).

7.4 Other Types of *t* Tests

Here we focused on a **single-sample *t* test**, but perhaps the most commonly used *t* test is a **two-sample *t* test**. A statistician would say that a two-sample *t* test is used when the null hypothesis is that two population means are the same, or that both samples were drawn from the same population. In simple language, this could mean that two treatments have the same effect, or that a treatment has no effect in comparison to a control, or that clams below a sewage treatment plant have the same average dry weight as those above (Section 2.3). The prediction that follows is that the two samples will have the same mean to an infinite number of decimal places. A two-sample *t* test can also be used to test for a difference other than zero, but normally there is no reason to test for a nonzero difference.

A *t* test for regression coefficient is used to compare the **slope of a line** with a hypothetical slope of zero. A statistician might state the null as *the population's slope is zero.* This could translate as *the amount of this has no relation to the amount of that* or *has no effect on the amount of that* . . . The prediction that follows is that the sample's line of best fit will have a slope of exactly zero (Section 2.3). Nonzero values for the population's slope can also be used, but normally we would not test for a nonzero slope.

There is a *t* test to **compare the slopes of two lines** to each other. A statistician's null might be that the samples were drawn from populations with the same slope, or that they were drawn from the same population. This could translate as *the amount of this has the same effect on the amount of that for both of these two conditions.* The prediction that follows is that the two samples' best fit lines will have exactly the same slopes (Section 2.3).

There is a *t* test for **paired data**, which is really a special case of a single sample *t* test (Section 13.1).

In all cases, the numerator in the formula for *t* is the difference between the prediction and the outcome, and the denominator is some form of standard error.

There are **analyses of variance** (ANOVAs) that are **mathematically equivalent** to *t* tests. The difference is that ANOVAs allow for the inclusion of larger numbers of samples. For example, a 1-way ANOVA allows the for the comparison of two or more samples while a two-sample *t* test is limited to two. If two samples are being compared, the two types of tests are equivalent: the ANOVA's test statistic, *F*, would be equal to t^2, and the two types of tests would yield the same *P* values. Similarly, analysis of covariance allows for the comparison of two or more slopes, while the equivalent *t* test is limited to two. Again, the tests are equivalent if only two samples are compared. We would find that $F = t^2$, and the two types of tests would yield the same *P*-values. In fact, ***t* tests are special cases of ANOVAs**, special cases because of the limited number of samples that can be analyzed with *t* tests.

7.5 ANOVAs and *t* Tests Have Certain Requirements

Analyses of variance, and *t* tests of course, require that the data have certain properties. The data should be drawn from populations with normal distributions. For our purposes, a **normal distribution** is a bell-shaped curve that is neither too flat nor too pointy. Most often we are comparing multiple samples to each other. In that case, the samples should be drawn from populations that have the same variance as each other (Section 7.2.1). Typically, it is appropriate to assume that both conditions are met and use *t* tests or ANOVAs. It is when data severely violate the requirements of those tests special action is required (Chapter 12).

7.6 Do Not Test for Equal Variances

Many make the mistake of testing the null hypothesis that variances are the same and using a *t* test or ANOVA if $P > \alpha$. In doing so, they are falling into the trap of trying to prove the null hypothesis (Section 3.3). A finding that $P > \alpha$ does not prove that the populations have the same variance.

7.7 Simple Principles Behind the Analysis of Counts of Observations Within Categories

7.7.1 Counts of Observations Within Categories

When it comes to counts of observations within categories, nominal data, we are generally interested in proportions, and we often use some type of chi-squared[1] test. We might want to know if two genes are on different chromosomes. If they are, then we would expect the outcome of our experiment to take the form of a 9 : 3 : 3 : 1 ratio. Chances are, we will not get that exact ratio, but can we rule out sampling error as the cause and conclude that the two genes are on the same chromosome? In cases like this, where the null dictates specific proportions, we analyze results with a **chi-squared goodness of fit test**.

[1] Or chi-square.

Instead, we might want to know how the proportions of Democrats, Republicans, and *other* differ when comparing Dade and Palm Beach Counties, in Florida. In that case, we have no idea what ratios to expect but, whatever they are, we use the null to predict that they are exactly the same for both counties. We know that such a null is almost certainly false (Section 2.5), but we want to know if we can exclude sampling error as the sole cause of the directional differences we see when comparing the two counties. In cases like this, in which the null does not specify a specific, numeric prediction, we use a **chi-squared test of independence**.

7.7.2 When the Null Specifies the Prediction

The following example illustrates the use of a **chi-squared goodness of fit test**. Consider one of Mendel's famous genetics experiments. Peas have a gene for color and a gene for shape. Are they on the same chromosome? The null hypothesis is that they are not because, if that were true, we could form a specific, numeric prediction. The outcome of our experiment should be a 9 : 3 : 3 : 1 ratio of peas that are yellow and round, yellow and wrinkled, green and round, and green and wrinkled. What we get is 30 : 12 : 6 : 3. That is close to 9 : 3 : 3 : 1, but it is not exactly 9 : 3 : 3 : 1. Can we exclude sampling error as the sole reason for that difference? We perform a chi-squared goodness of fit test. Our results are the *observed values* (or *frequencies*) (Table 7.3). The expected values are the predictions, given the sample size. In other words, the expected values must sum to the same value as the sum of the observed values, in this case, 51 peas. Since we expect a 9 : 3: 3 : 1 ratio, we calculate the expected values as 9/16 of 51, 3/16 of the 51, and so on.

Table 7.3. Observed and expected values for a goodness of fit test

	Y/R	Y/Wr	G/R	G/Wr
Observed	30	12	6	3
Expected	28.69	9.56	9.56	3.19

Even though there cannot be a fraction of a pea, expected values should not be rounded to whole numbers. The more they are rounded, the more we miscalculate the test statistic, X^2. Here I rounded to two decimal places for convenience.

Next, we subtract expected from observed, square the result, and divide by expected, for each of the four categories. Then we sum the outcomes.

$$X^2 = \frac{(\mathrm{O}-\mathrm{E})^2}{E} + \frac{(\mathrm{O}-\mathrm{E})^2}{E} + \frac{(\mathrm{O}-\mathrm{E})^2}{E} + \frac{(\mathrm{O}-\mathrm{E})^2}{E}$$

$$X^2 = 2.02$$

Next, we need degrees of freedom (DF), which is the number of categories minus one. There are four categories in this example, the four kinds of peas (expected values do not qualify as categories), so there are three degrees of freedom. We estimate P the same way we did with the t test, but with a table for chi-squared (Table 7.4).

Table 7.4 A probability table for X^2

	P (2-tailed)					
DF	0.50	0.30	0.20	0.10	0.05	0.01
1	0.46	1.07	1.64	2.71	3.84	6.64
2	1.39	2.41	3.22	4.60	5.99	9.21
3	2.37	3.66	4.64	6.25	7.82	11.30
4	3.36	4.88	5.99	7.78	9.49	13.30

Our X^2-value of 2.02 is to the left of 2.37, so $P > 0.50$. We cannot safely exclude sampling error for the difference between the prediction and the outcome. In this example, the null could be correct, and we are not interested in the direction of the outcome in comparison to the prediction. The question is whether to reject or fail to reject. We fail to reject and draw no conclusion.

7.7.3 When the Null Does Not Specify the Prediction

We use a chi-squared test of independence when the null cannot be used to directly to determine expected values. We may want to compare party affiliation of voters in Dade County, Florida, with the affiliations of those in Palm Beach County. Our null hypothesis is that the variables *party affiliation* and *county* are independent of each other. This could be rephrased as *party affiliation is the same in both counties.* We have no knowledge of how many Democrats, Republicans, and *other* there are in those two counties. We cannot calculate expected values the way we do for goodness-of-fit. The null does not dictate specific proportions. Nevertheless, our null hypothesis allows for a prediction, but we need to collect the data to formulate it. Here is how it is done. Imagine that we poll 195 voters and get the results shown in Table 7.5.

Note the totals. There were 140 Democrats, 16 Republicans, and 39 *other*, for a total of 195 voters. So, Democrats make up 140/195, or 71.8%, of the total. Ninety-five persons from Palm Beach County provided party affiliation. If county and party affiliation are independent of each other, then 71.8% of the 95 persons in Palm Beach County should be Democrats, 68.21 persons, which is the expected value for that cell.

Table 7.5 Data for a test of independence

	Observed values			
County	Democrats	Republicans	Other	Total
Dade County	65	14	21	100
Palm Beach Cty.	75	2	18	95
Total	140	16	39	195

We can also get the expected value for that cell the other way around. Persons in Palm Beach County make up 95/195, or 48.7%, of the total number of persons who gave their party affiliation. The total number of Democrats is 140. If the null hypothesis is correct, we expect 48.7% of 140 to be the number of Democrats in Palm Beach County. Again, we get 68.21. We do the same for the other five cells. The result is referred to as a contingency table (Table 7.6). Expected values are shown italicized and in parentheses.

Table 7.6 A contingency table for a test of independence

County	Democrats	Republicans	Other	Total
Dade County	65	14	21	100
	(71.79)	*(8.21)*	*(20)*	
Palm Beach Cty.	75	2	18	95
	(68.21)	*(7.79)*	*(19)*	
Total	140	16	39	195

The calculation of X^2 is as in Section 7.7.2, with observed minus expected, squared, over expected *for all six cells* i.e., for every combination of party affiliation and county. We sum the results for X^2.

$$X^2 = 9.81$$

Rounding the expected values had a trivial effect. Without rounding, the result would have been 9.82.

In a test of independence, the number of degrees of freedom is equal to the number of rows in the contingency table, minus one, times the number of columns, minus one. In this example,

$$\text{DF} = (r - 1)(c - 1)$$
$$\text{DF} = (2 - 1)(3 - 1)$$
$$\text{DF} = 2.$$

7.8 Interpreting *P* When the Null Hypothesis Cannot Be Correct

Continuing from Section 7.7.3, we consult Table 7.4 and find that $P<0.01$. In this situation, it is nearly impossible for the null to be true (Section 2.5). We want to know if we can exclude sampling error as the cause of the differences in proportions that we found. We can, but for which differences? That question is addressed in Section 7.10.

7.9 2 × 2 Designs and Other Variations

A goodness of fit test requires at least two categories. When we asked if two genes are on different chromosomes (Section 7.7.2), we had four. Similarly, a contingency table can have various numbers of rows and columns. The smallest contingency table would have just two of each, a 2 × 2 table. When working with a 2 × 2 contingency table, an alternative to a chi-squared test of independence is **Fisher's exact test**, and Fisher's exact test is the best choice when sample sizes are low. The mathematics would be found in any "proper statistics book", but no one performs Fisher's exact test by hand.

In addition to allowing for various numbers of rows and columns, contingency tables can accommodate more than two variables, which would add more dimensions to the table. Any "proper statistics book" would show the procedure for performing the test of independence, but the results would be hard to interpret. This is because there is a problem with chi-squared tests (Section7.10).

7.10 The Problem with Chi-squared Tests

Consider the example of party affiliation in two Florida counties (Section 7.7). The *P*-value tells us we can exclude sampling error as the cause, but for which differences? This is the problem with chi-squared tests, both goodness of fit tests and tests of independence. If there is more than one degree of freedom, there is more than one comparison being made. A *P*-value may tell us that we can exclude sampling error as the sole cause of a difference but, if there is more than one difference, for which ones? It may be that Dade County has a lower proportion of Democrats than Palm Beach County, a higher proportion of Republicans, a higher proportion of *other*, or any combination of the three.

There is no good solution to this problem. On approach would be to construct three more contingency tables and calculate X^2 for each of them. One table would include Democrats and Republicans, another Democrats and *other*, and the last Republicans and *other*. The problem with this approach is that we are still testing the same null hypothesis, and we are testing it a total of four times. Anytime we test a null, we take that 5% risk that we will exclude sampling error when sampling error alone is the cause of a difference, i.e., make a type I error (Section 2.6). By testing one null four times we elevate that risk well above 5% (Section 9.1). We could say that the follow up contingency tables are justified by the initial finding that $P < \alpha$, but the three follow up contingency tables carry a "familywise" type I error rate of 14%.[1] A similar issue occurs with ANOVA but, in that case, there are specific post hoc tests that can be used to keep the familywise type I error rate low (Section 9.3).

It might be tempting to perform just one more contingency table analysis and base it on what looks like the most meaningful finding. In this case, we might choose to compare just Democrats with Republicans. Now the problem is that we are examining our data, formulating a new hypothesis, and testing it with the same data that inspired that hypothesis. That solution produces an invalid result (Section 2.8).

If it is important to tease apart the results of goodness of fit tests or tests of independence in which there is more than one degree of freedom, we must consult Sharpe.[2] He provides four approaches, all of which are serious undertakings. He also provides the following, surprising advice. "If you can avoid chi-square contingency tables with greater than one degree of freedom, you should do so. For example, a researcher might collapse or discard low frequency cells after collecting the data but prior to conducting a chi-square test." In some cases, that might be good advice when it comes to goodness of fit tests too.

7.11 The Reasoning Behind the Mathematics

Recall that, for numerical data, many statistical tests combine the difference between the prediction and the outcome, the amount of scatter in the data, and sample size into one test statistic. With nominal data, there is no scatter, so the logic behind chi-squared tests is to combine the difference between the prediction and the outcome with sample size into X^2. The difference between the prediction and the outcome is squared in the numerators as we calculate X^2. So, the greater the discrepancy between the prediction and the outcome, the larger the value of X^2. Meanwhile, sample size is also built into the calculation of X^2. Larger sample sizes generate larger X^2 values. Finally, as is the case with t, the greater the value of X^2, the lower the P-value.

7.12 Rules for Chi-squared Tests

The following are rules for chi-squared tests, both for goodness of fit tests and tests of independence.

---

[1] For why it is not 15%, see J. Zar. 2010. Biostatistical Analysis, 5th ed. Prentice Hall, p. 189.

[2] D. Sharpe. 2015. Your chi-square test is statistically significant: now what? Practical Assessment, Research & Evaluation, 20: article 8.

- Do not take the square root of X^2. The exponent is part of the symbol.
- Actual counts must be used for observed values. Converting them to percentages creates an incorrect sample size of 100. Converting them to a range zero to one incorrectly sets the sample size to one.
- Expected values must be calculated so that they sum to the sum of the observed values. They are what we expect given the sample size.
- Under certain circumstances, expected values should not be less than five. If we obtain an expected value less than five, we should consider dropping the category, combining it with another, or increasing the sample size. Zar[1] describes the circumstances in which it is acceptable for an expected value to be less than five.

[1] J. Zar. 2010. Biostatisical Analysis, 5th ed. Prentice Hall, pp. 473,474.

CHAPTER 8
THE TWO-SAMPLE *T* TEST AND THE IMPORTANCE OF POOLED VARIANCE

All *t* tests combine three aspects of the data into one test statistic: the difference between the prediction and the outcome, the sample sizes, and the amount of scatter in the data. If there are two samples, the test is conducted much like a single sample *t* test (Section 7.2), except that we have to account for presence of two samples. We calculate variance based on both samples combined, and we use a different numerator in the formula for *t*. We would be conducting a **two-sample *t* test**. Because a two-sample *t* test is one of the few tests we are likely to perform by hand, and because of the importance of pooled variance, which will be introduced, it is worth it for us to be familiar with the mathematics of a two-sample *t* test.

Our example will be the dry weight of clams found below a sewage treatment plant and above. There is no research hypothesis (Section 2.9). Clams could be heavier below the plant than above, or it could be the other way around. The null is that the average dry weight of the clams below the plant is the same as above, to an infinite number of decimal places. That null cannot be true, but it allows for a specific, numeric prediction: the two samples will have exactly the same average dry weight.

We randomly choose ten clams from each location and perform the same initial calculations as in a single sample test (Example 8.1, subscripts indicate which sample is which). Recall variance, the average squared deviation from the mean, a commonly used index of scatter (Section 7.2.2). When there are two samples, we must calculate **pooled variance**, (***error MS***). Recall that variance often reflects natural variation, even though it is referred to as error. The clams vary in mass.

One of the requirements of a *t* test, or ANOVA, is that the populations have the same variance (Section 7.5). We generally assume that is the case, so it is safe to pool the variances from both groups. It provides the best estimate of the populations' variance. Pooled variance is a very important statistic. It plays a key role in ANOVA (Section 9.2).

Example 8.1 Dry weights of clams collected above and below a sewage treatment plant and the two-sample *t* test.

Dry weight of clams upstream (g)	$(x_1 - \bar{x}_1)^2$	Dry weight of clams downstream (g)	$(x_2 - \bar{x}_2)^2$
2.16	12.62	6.22	1.72
6.52	0.65	6.96	0.33
5.57	0.02	5.77	3.11
3.77	3.78	10.22	7.22
6.91	1.43	9.32	3.19
3.19	6.37	8.50	0.94
7.96	5.05	7.91	0.14
8.15	5.94	3.41	17.00
5.90	0.03	8.72	1.41
7.00	1.66	8.30	0.59
$\bar{x}_1 =$ 5.71		$\bar{x}_2 =$ 7.53	
$SS_1 =$	37.55	$SS_2 =$	35.65
$n_1 =$ 10		$n_2 =$ 10	
$df_1 =$ 9		$df_2 =$ 9	

Pooled variance is calculated as follows:

$$error\ MS = \frac{(SS_1 + SS_2)}{(df_1 + df_2)}$$

$$error\ MS = 4.07$$

The next step is to calculate **standard error of the difference between the means, $S_{(\bar{x}1-\bar{x}2)}$**. This is a different quantity from standard error of the mean (Section 7.2.1). Like standard error of the mean, standard error of the difference between the means factors in variance and sample size. Standard error of the difference between the means is calculated as follows.

$$S_{\bar{x}1-\bar{x}2} = \sqrt{\frac{error\ MS}{n_1} + \frac{error\ MS}{n_2}}$$

$$S_{\bar{x}1-\bar{x}2} = 0.902$$

Note that a large amount of scatter in the data, or a small sample size, favor large values of standard error of the difference between the means. Large values of standard error suggest that we should not be confident in drawing conclusions from our data. The opposite is true if standard error is low.

The last step is to factor in the difference between the prediction and the outcome. The prediction is that the sample means will be exactly the same. The result is *t*, a number that combines all three aspects of the data into one quantity. The following shows how *t* is calculated. The symbol μ_0 is the predicted difference between the means. Since this usually zero, as in this case, μ_0 is sometimes omitted from the formula.

$$t = \frac{|\bar{x}_1 - \bar{x}_2| - \mu_0}{S_{(\bar{x}_1 - \bar{x}_2)}}$$

$$t = 2.02$$

Recall that a low value of standard error suggests that we can be confident in the data. A low value of standard error would tend to make *t* large. The value of *t* would also be large if there were a large difference between the means, i.e., a large discrepancy between the prediction and the outcome. So, *t* is a combined representation of sample size, variation, and the difference between the prediction and the outcome. A high value of *t* suggests that we should be confident in our data, a low value the opposite.

Lastly, we need to determine *P*. For that we need *total* degrees of freedom, *DF* which, for a two-sample *t* test, is the sum of df_1 and df_2, in this case $9 + 9 = 18 = DF$ (cf. Section 7.2.2). We locate our value of *DF* in the left-most column of the *t* table (Table 8.1), read across to the right to approximate the location of *t*, then read up to the top row for the *P*-value.

Table 8.1 A probability table for *t*

	P(2-tailed)					
DF	0.50	0.20	0.10	0.05	0.01	0.001
17	0.689	1.333	1.740	2.110	2.898	3.965
18	0.688	1.330	1.734	2.101	2.878	3.922
19	0.688	1.328	1.729	2.093	2.861	3.883

Our *t* value of 2.02 is between 1.734 and 2.101, so *P* is less than 0.10 and greater than 0.05. If we adopt the traditional criterion of $\alpha=0.05$, then $P>\alpha$. We draw no conclusions. But *P* is close to α. We might be able to draw conclusions if we collected just a few more clams. We know the null hypothesis must be false. It is not possible that the mean mass of the clams above the plant is the same as below it to an infinite number of decimal places. By collecting more clams, we might be able to determine whether they are heavier below the treatment plant or above. If we did, we would *not* be able to generalize to locations near other sewage treatment plants (Section 1.5)

CHAPTER 9
COMPARING MORE THAN TWO GROUPS TO EACH OTHER: ANALYSIS OF VARIANCE

9.1 If We Have Three or More Samples, We Cannot Use Two-Sample *t* Tests to Compare Them Two Samples at a Time

Suppose we have treatments A, B, and C. Those treatments yield samples A, B, and C. The sample means are all different, but we need to determine if any of those differences are meaningful. We use two-sample *t* tests to compare samples A with B, A with C, and B with C. Wrong. The problem is that there is one null hypothesis, and we are testing it three times. A statistician might state the null as *all three samples were drawn from the same population.* In practice, the null might be *all three treatments have the same effect* or, if one is a control, *they have no effect.* Every time we test a null, we accept that 5% risk but, if we test one null three times, it is not 5%. It is 14%.[1] That 14% is termed **familywise type I error**. Obviously, the more nulls we test, the more opportunities there are to draw the wrong conclusions, but there are ways to limit that problem. One of those ways is to avoid testing the same null more than once. That can be done with **analysis of variance**, **ANOVA**, one of the many great gifts of Ronald Fisher.

9.2 Simple Principles Behind the Mathematics

Analysis of variance works on the same sorts of data as *t* tests, but ANOVAs allow for the inclusion of more than two samples. All samples are compared to each other simultaneously in a single test. This holds familywise error to 0.05. How does ANOVA work? Simple.

Imagine that we manufacture dog food. We develop a new recipe, and we want to compare it to our old one, and those of other brands. We will call the new dog food New

[1] For why it is not 15%, see J. Zar. 2010. Biostatistical Analysis, 5th ed. Prentice Hall, p. 189,190.

Chow, the old one Old Chow, and the other three competitor 1, competitor 2, and competitor 3; C1, C2, and C3.

We assign ten dogs to each of the dog foods and see how much dog food they eat per day. To analyze the results, we use a **1-way ANOVA**, 1-way because there is one independent variable, *dog food*. The null hypothesis is that dogs have no preference among any of the chows, but a statistician might say that all samples were drawn from the same population. The statistician's null allows us to understand ANOVA. If we sample the same population multiple times, not only can we estimate the population's variance as *pooled variance* (*error MS*; Section 7.2), we can also estimate it with the use of the different sample means (*groups MS*). Any "proper statistics book" would show how. Since both are estimates of the same quantity, the population variance, if we divide one by the other, we should get one. In ANOVA, to divide one by the other, we create an F-ratio.[1]

$$F = \frac{groups\ MS}{error\ MS}$$

The precise, numeric prediction based on the null is that $F = 1$ to an infinite number of decimal places.

If, however, the samples were drawn from populations with different means, then the sample means would be farther apart from each other, *groups MS* would overestimate the population variance, and $F > 1$. In ANOVA, if F is sufficiently greater than one, we conclude that our results are meaningful. There is too great a difference between the outcome and the prediction for the samples to have been drawn from the same population.

While the mathematics of ANOVA appears different from that of a two-sample t test, recall that the latter is really a special case of the former (Section 7.4). What this means is that ANOVA combines the same three aspects of the data into its statistic, F, that a t test does into its statistic. Those characteristics are the amount of scatter in the data, the sample sizes, and the difference between the prediction and the outcome. High F-ratios and low P-values are favored by little scatter in the data, large sample sizes, and large differences between the prediction and the outcome.

Suppose we get the results in Figure 9.1. In reality, Figure 9.1 shows five samples from the same population. There is no preference for any dog food. The population mean is 50 g of dog food per day, and the population variance is 225. To perform a 1-way ANOVA, we calculate pooled variance as in Chapter 8, but with five sums of squares in the numerator and five degrees of freedom in the denominator. The result is *error MS* = 258.6. We use the five sample means to calculate *groups MS* and obtain the

---

[1] Named for Fisher.

value 230.1. Since both are estimates of the same population variance, 225, values of *error MS* and *groups MS* are similar to each other.

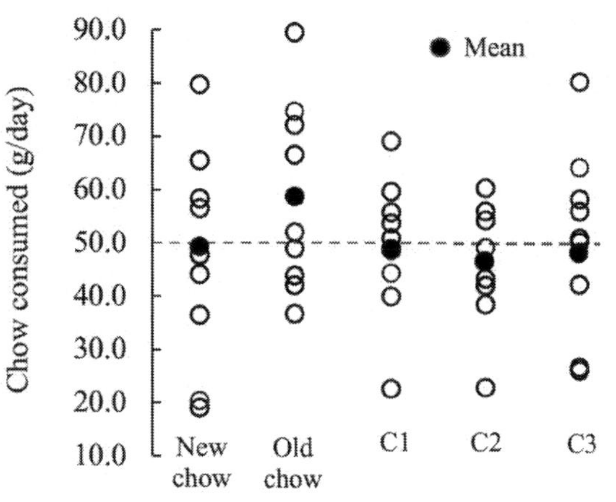

Figure 9.1 Five random samples from a population with a mean of 50, a variance of 225, and a standard deviation of 15.

Consequently,

$$F_{4,45} = \frac{230.1}{258.6}$$

$$F_{4,45} = 0.89$$

with 4 and 45 being the degrees of freedom associated with the numerator and denominator. If $F_{4,45} = 0.89$, then $P = 0.48$. We draw no conclusion.

Suppose we get the results in Figure 9.2 instead. In this case, there is a strong preference for New Chow. Dogs, in general, would eat 75 g/day of it, that value being the population mean. If the corresponding sample mean is used along with the others to

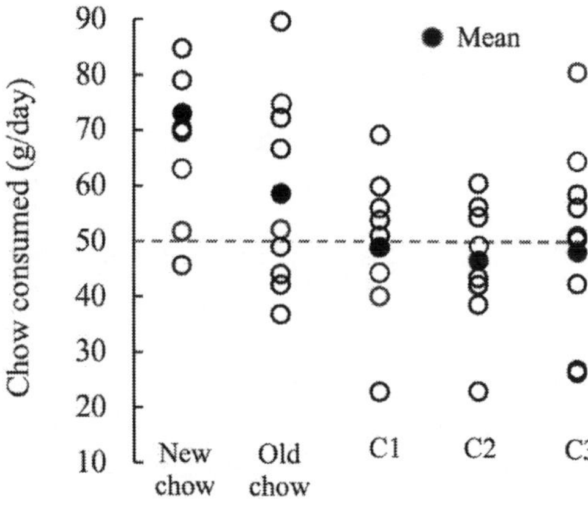

Figure 9.2 Data shown are the same as in Figure 9.1, except that, for New Chow, the sample was drawn from a population with a mean of 75.

estimate variance, the result will be an overestimate, 1224.

We perform a 1-way ANOVA, and get the following.

$$F_{4,45} = \frac{1224.0}{234.4}$$

$$F_{4,45} = 5.22$$

$$P = 0.002$$

A statistician would conclude that the samples were drawn from different populations. We would conclude that there is a difference in preference and, by comparing all samples to each other simultaneously, we kept α at 0.05

9.3 Follow Up Tests Conducted After the ANOVA

Continuing from Section 9.2, we found that there is a preference, but for which dog foods over which? The next step is to follow up with specific **post hoc tests** that maintain a familywise error rate of α.[1] Two-sample *t* tests would not do the same. There is a widespread belief that post hoc tests should not be performed without first performing an ANOVA and finding that $P \leq \alpha$. In fact, it is valid to perform good post hoc tests, like the Tukey honestly significantly different test, without an initial ANOVA finding of $P \leq \alpha$,[2] but so many people would think the tests unjustified, that I always run the ANOVA first, and perform post hocs if the result is that $P \leq \alpha$.

Some circumstances (Sections 10.3 and 13.3) require that post hoc tests be modified. Since these modified versions are not conducted by all popular software packages, it is important to be able to perform post hoc tests by hand, so we can modify them when needed. To perform a Tukey test, we first calculate standard error (SE) as follows

$$SE = \sqrt{\frac{error\ MS}{n}}$$

in which *n* is the sample size of each group, and *error MS* is the denominator in the *F*-ratio for the ANOVA. In this case,

---

[1] Fisher's least significant difference test should not be used, because it does not adequately control familywise error.

[2] J. Zar. 2010. Biostatistical Analysis. Prentice Hall, p. 226.

$$SE = \sqrt{\frac{234.4}{10}}$$

$$SE = 4.84.$$

If sample sizes differ, standard error can be calculated as follows

$$SE = \sqrt{\left(\frac{error\ MS}{2}\right)\left(\frac{1}{n_1} + \frac{1}{n_2}\right)}$$

in which n_1 and n_2 are the sample sizes of the two groups being compared. When this formula for is used for standard error, the test is a **Tukey-Kramer** test.

$$SE = \sqrt{\left(\frac{234.4}{2}\right)\left(\frac{1}{10} + \frac{1}{10}\right)}$$

$$SE = 4.84.$$

Next, we calculate q, which is the statistic produced by the Tukey test.

$$q = \frac{|\bar{x}_{\mathrm{A}} - \bar{x}_{\mathrm{B}}|}{SE}$$

The numerator is the absolute value of the difference between the two sample means that are being compared. In this case, there are five sample means, so each would be compared to each other, for a total of ten Tukey tests. Each would constitute a **pairwise comparison**, the pairs being the two samples being evaluated.[1] For example, if we compare New Chow to Old Chow,

$$q = \frac{|72.8 - 58.6|}{4.84}$$

$$q = 2.93.$$

We use the degrees of freedom associated with *groups MS* (Section 9.2) along with the total number of groups being compared, k, to determine P with a probability table. The degrees of freedom are 45 and the number of groups is five. Our probability table may not show our exact degree of freedom, so we use the table's nearest lower degree of freedom. For example, my table shows 40 and 50 for the degrees of freedom associated

---

[1] For arguments against making all possible pairwise comparisons see Wilkinson and the Task Force on Statistical Inference.1999. American Psychologist, 54(8): 594-604.

with *error MS*, so we choose 40. For $P = 0.05$, we find that q must be 4.039 or greater to draw a conclusion. Although the dogs, on average, preferred New Chow to Old Chow, we cannot exclude sampling error as the cause of that directional difference.

Results of all the Tukey tests indicate that $P < \alpha$ for the three comparisons of New Chow with the three competitors. These results are used in Section 14.5 to illustrate the best way to plot the results of post hoc tests.

9.3.1 When Comparing Multiple Groups to a Single Reference, Like a Control

Why did we conduct ten pairwise comparisons in the previous section?[1] What we really want to know is how New Chow stands up in comparison to the others. We only needed to conduct four tests, not ten. New Chow has special status. It is a **reference group**. Often, a reference group is a control group,[2] but a reference group is any group with special status. If we are comparing samples to a single reference, such as a control, and not to each other, we are not performing as many pairwise comparisons as we would otherwise. This means we can use a more liberal test than a Tukey test and still keep familywise error at α. A more liberal test would give more power (Section 2.7). **Dunnett's test** is such a test. The procedure can be found in any "proper statistics book".

Dunnett also recommends that the **reference group** has the **largest sample size**. Here is why. Suppose we were limited to 40 observations—perhaps we can only house 40 mice. Imagine that we have one reference group and three treatment groups, so we divide the mice up equally into four sets of 10. Each comparison of a control to a treatment would involve 20 data, 10 replicates from each sample. Imagine instead that the reference group has 16 data and the other three groups only 8. The total number of replicates is still 40, but each pairwise comparison would involve 24 data rather than 20, 16 from the reference and 8 from the treatment group. The greater the sample size, the greater the power, so 16 and 8 is better than 10 and 10. Often, we think, *this is just the control*, but a reference group is the most important group of all. If groups are only being compared to a reference, and not to each other, the reference should have the largest sample size, and Dunnett's test should be used.

Dunnett's paper is very dense.[3] According to Zar, the sample size for the reference group should be a little less than n_r, which is calculated as follows[4]

---

[1] Some authors argue against making all possible pair-wise comparisons, e.g. L. Wilkinson and the Task Force on Statistical Inference. 1999. American Psychologist, 54(8): 594–604.

[2] In a narrow sense, a control group is a comparison group that does not receive a treatment.

[3] C. Dunnett. 1955. Journal of the American Statistical Association, 50: 1096–1121.

[4] J. Zar. 2010. Biostatistical Analysis 5th ed. Prentice Hall. p. 235. This is my version of Zar's relationship. He uses more prose.

$$n_r = n_o\sqrt{\#other}$$

in which n_o is the sample size of the other groups and *#other* is the number of groups being compared to the reference. For example, if there are three groups being compared to the reference, and eight observations in each, the best sample size for the reference group would be a little lower than

$$13.9 = 8\sqrt{3}.$$

If there were 13 in the reference group and eight in the treatment groups, the total would be 37. Above, we had a total of 40. Would we get closer to 40 with 9 in each treatment group?

$$15.6 = 9\sqrt{3}$$

With 15 in the reference and 9 in the others, the total would be 42. We want to use all 40 mice, so we would have to delve into Dunnett's dense paper to find the solution.

CHAPTER 10
ASSESSING THE COMBINED EFFECTS OF MULTIPLE INDEPENDENT VARIABLES: THE JEWEL OF NULL HYPOTHESIS TESTING

10.1 Independent Variables Alone and in Combination

Sometimes it seems like null hypothesis testing only tells us what we already know. If a pattern is obviously not the result of sampling error, we know that *P* will be low. If a pattern could easily be the result of sampling error, we know that *P* will be high. If data are ambiguous, we know that *P* will be near α, and we should be cautious if we draw any conclusions. Where null hypothesis testing is most valuable is when it is used to assess the combined effects of multiple independent variables. By testing nulls, we can determine **how the effect of one independent variable depends on another independent variable**. As Joe Biden might say, **this is a big f&^%ing deal**. We often cannot assess the effect of one independent variable on another independent variable by just looking at the data. The ability to accomplish this is of **huge importance**, and we can do it with another type of analysis of variance, a multiway ANOVA. In this section, there will be two independent variables, so we will see the results of **2-way ANOVAs** (or 2-factor).

Imagine we develop a new type of car tire. We want to know if it lasts longer than our currently marketed tire, and we want to know if matters whether the tire is on the front or the rear of the car. The dependent variable is tire *longevity*, and there are two independent variables, *tire* and *position*. When there are multiple independent variables, we generally call each one a **factor**. The variations within them are termed **levels**. *Currently marketed* and *new* are the levels within the factor *tire*. *Front* and *rear* are the levels within the factor *position*. We conduct our experiment with ten identical cars in each combination of the independent variables. When all combinations of the independent variables are represented, the design is said to be **factorial**. Factorial designs are very common, and they must be analyzed with ANOVA. Factorial designs are often shown as matrices. Each combination of the independent variables is shown as a *cell*, boxed in Figure 10.1.

Figure 10.1 A factorial design with cell and marginal means. Each combination of tire and position is represented.

	Front	Rear	Main effect of tire
Currently marketed	$n_{f,cm} = 10$; Cell mean, $\overline{x}_{f,cm}$	$n_{r,cm} = 10$; Cell mean, $\overline{x}_{r,cm}$	$n_{cm} = 20$; Marginal mean, $\overline{x}_{cm}$
New	$n_{f,n} = 10$; Cell mean, $\overline{x}_{f,n}$	$n_{r,n} = 10$; Cell mean, $\overline{x}_{r,n}$	$n_n = 20$; Marginal mean, $\overline{x}_n$
Main effect of position	$n_f = 20$; Marginal mean, $\overline{x}_f$	$n_r = 20$; Marginal mean, $\overline{x}_r$	

Each cell would contain a sample, which we could use to calculate a sample mean or **cell mean**. But we could also pool results across the cells to look at one variable at a time. For example, if we were only interested in the variable *tire*, we could pool results horizontally, combining the results from the two different positions. Those pooled results would go in the right margin, outside of the matrix. From those pooled results, we could calculate **marginal means**. Similarly, marginal results at the bottom would pertain to *position*.

Since the design entails two independent variables, we analyze the results with a 2-way ANOVA, which tests **three null hypotheses**. The outcome will be *F*-ratios and *P*-values for all three. Most often the nulls are that one independent variable has no effect, the other has no effect, and the effect of one independent variable does not depend on the other one. In this case, the nulls would be that both tires have the same longevity, that tire longevity is not dependent on position, and that any effect of position on longevity is the same for both kinds of tires. The latter could be rephrased as *any difference in longevity between the tires does not depend on the tire's position.*

For the first two nulls, we pool the results across one independent variable to analyze the other, i.e., we analyze the results in the margins, not the cells. If $P \leq \alpha$, then there is a **main effect** of that variable. If we consider the results in the right margin, and we find that $P \leq \alpha$, there would be a main effect of *tire*. Since the different tires have different compositions and tread patterns, the null cannot possibly be correct—one must wear out faster than the other (Section 2.5). A main effect of *tire* means that we can confidently say that whichever tire wore out faster in our study is the tire that wears out faster in general. Similarly, the results in the bottom margin could be used to test for a main effect of position, since results are pooled from the two types of tires. Again, we have a null that cannot be true. Position must have *some* influence on tire wear. Finally, we would use the results in the cells to determine if position affects which tire wears out faster. That is the same as asking if the type of tire influences the effect of position on longevity. When considering the combined effects of multiple independent variables, if $P \leq \alpha$, there is said to be an **interaction** between the variables. With such an outcome, **we can compare differences**, as will be illustrated.

The following examples should help us understand main effects and interactions. For the sake of argument, assume that $P \leq \alpha$ for all differences, i.e., none are the product of sampling error alone.

Figure 10.2 shows no main effects or interaction. We cannot draw conclusions. In comparison, there is a main effect of *position* in Figure 10.3. Front tires wear out faster than rear tires. There is no main effect of *tire* and no interaction. We cannot tell which type of tire wears out faster, and how its faster wear is influenced by its position.

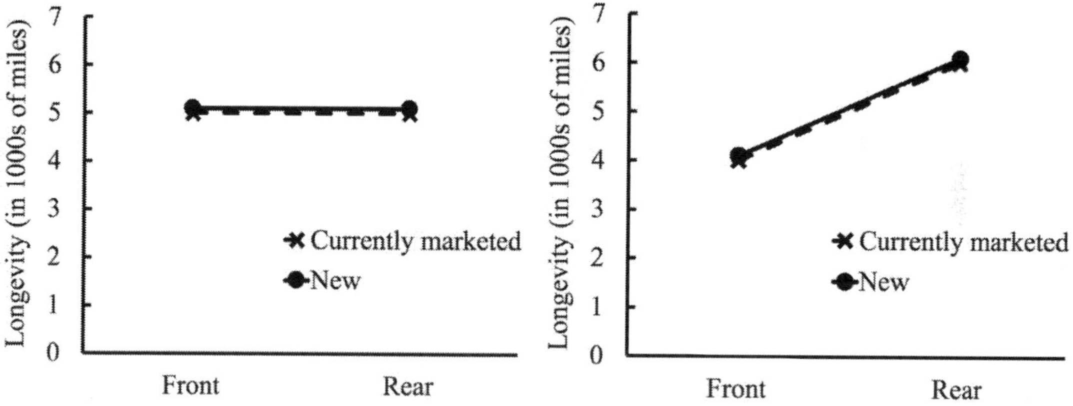

Figure 10.2 No main effects or interaction.

Figure 10.3 A main effect of position.

Figure 10.4 shows a main effect of *tire*. The new tires last longer than the ones that are currently marketed. There is no main effect of position and no interaction. We cannot say how position affects tire longevity, and we cannot say how position might affect the two types of tires differently.

There are no main effects in Figure 10.5. How do we know that? Consider *position*. Currently marketed tires went 60,000 mi in the front position, and the new ones went 40,000 mi, for a marginal mean of 50,000 mi in the front. In the rear it was the other way around, but the marginal mean is the same, 50,000 mi. So, no main effect of *position*. Similarly, if we lump together all the results for the currently marketed tire, we get a marginal mean of 50,000 mi, and we get the same number if we lump together all the results for the new one. Again, no main effect.

But we have an interaction. The new tire wore out quickly in the front and went a long time in the rear. In the currently marketed tire, it was the other way around. The differences have the same magnitude, 20,000 mi, but the directions of those differences are opposite. **The interaction tells us that the different directions of those differences are meaningful**.

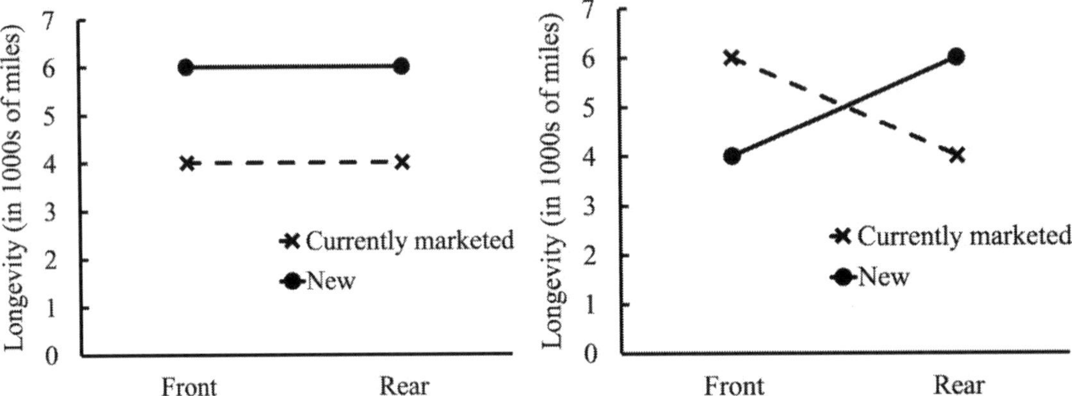

Figure 10.4 A main effect of tire.

Figure 10.5 An interaction between tire and position

Figure 10.6 shows an interaction. Longevity differs more in the front than in the rear. In the front, the difference is 20,000 mi. In the rear, the sample means hardly differ. **Given the interaction, that difference in differences is meaningful**. In general, there is a greater difference when the tires are in the front than in the rear. This can still be considered a question of direction (Section 2.5). Thanks to the ANOVA, we can say that the difference is greater at the front than at the rear, and not the other way around.

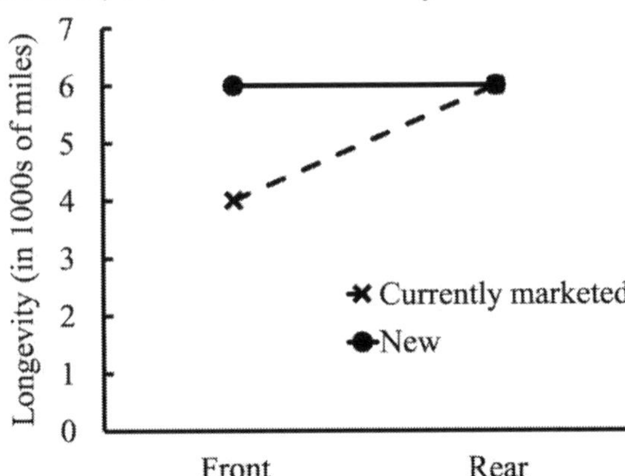

Figure 10.6 An interaction that creates a meaningless main effect.

So, interactions can tell us that differences in direction are meaningful, as in Figure 10.5, and that differences in magnitude are meaningful, as in Figure 10.6.

There is more. Figure 10.6 also shows main effects and, in this case, they are created entirely by the interaction, so they are unimportant by themselves. To see why, we need to consider marginal means. Marginal means at the bottom of Figure 10.7 show the main effect of position. On average, the tires lasted 50,000 mi in the front and went 60,000 mi in the rear. But this main effect was created entirely by the currently marketed tire, which only lasted 40,000 mi in the front. The new tires faired about as well in the front as in the rear.

Figure 10.7 Marginal means illustrate how an interaction can create a main effect that is unimportant.

	Front	Rear	Main effect of tire
Currently marketed	$\overline{x}_{f,cm}$ = 40,000 mi	$\overline{x}_{r,cm}$ = 60,000 mi	$\overline{x}_{cm}$ = 50,000 mi
New	$\overline{x}_{f,n}$ = 60,000 mi	$\overline{x}_{r,n}$ = 60,000 mi	$\overline{x}_{n}$ = 60,000 mi
Main effect of position	$\overline{x}_{f}$ = 50,000 mi	$\overline{x}_{r}$ = 60,000 mi	

We also have a main effect of *tire*. The right margin shows that the new tires last about 60,000 mi, which is 10,000 mi longer than the currently marketed tires. But this is only because the new ones last much longer in the front. These results illustrate how **an interaction can create main effects that are unimportant in themselves**.

In comparison, Figure 10.8 shows an interaction and main effects but, in this case, **the main effects are important**. The new tires last longer than the old ones in both positions. Also, the new tires do especially well up front, where they went 20,000 mi longer than the currently marketed ones. In the rear, they only went 10,000 mi longer. The interaction tells us that the difference between those differences, 10,000 and 20,000, is meaningful.

Multifactor ANOVA is perhaps Fisher's greatest gift. He taught us when to exclude sampling error as the sole cause of differences in magnitude or direction. Without ANOVA, we would never be able to assess the effects of multiple independent variables. Apparently, those who believe that null hypothesis testing should be abandoned never work with more than one independent variable.

Figure 10.8 Both the interaction and the main effect are important.

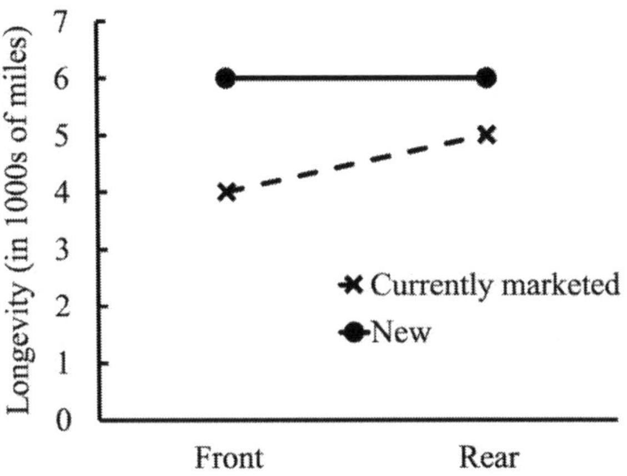

10.2 No, We May Not Use Multiple *t* Tests

Suppose we get the results shown in Figure 10.8, but we do not perform a 2-way ANOVA. Instead, we compare the two types of tires at the front, with a two-sample *t* test, and the two types of tires in the rear, with another *t* test. It is popular to analyze data this way, and it is **completely invalid**. To see why, imagine we get the results in Figure 10.9.

We decide that the new tire does better than the currently marketed one at the front, but not at the rear. Wrong. Here is why.

- Saying the tires do not differ at the rear constitutes accepting the null hypothesis. Instead, we should form no conclusion (Section 3.3).
- When it comes to an interaction, there is a single null hypothesis, though it can be stated in two ways. Here the null is that any difference in longevity between the two tires would be unaffected by *position*. We could also say that any effect of *position* is the same for both tires. Either way, it is the same null hypothesis, so it should only be tested once. ANOVA does that. Use of two *t* tests does not. Consequently, familywise error is above α. We are giving ourselves two opportunities to draw an incorrect conclusion instead of one (Section 9.1).

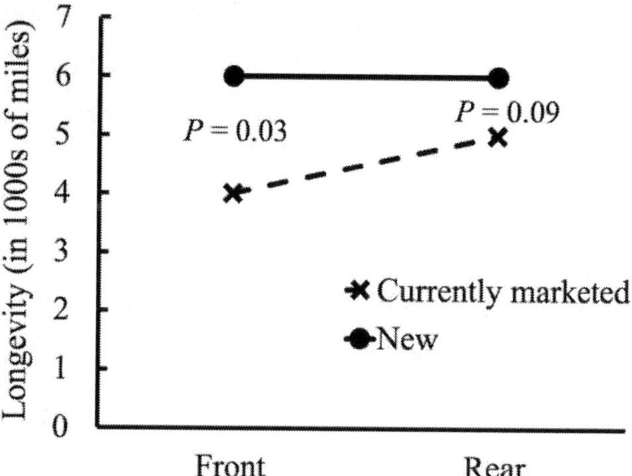

Figure 10.9 Results of performing two tests instead of a single test for an interaction. It is invalid to perform two tests instead of one in cases like this.

Longevity (in 1000s of miles)
7
6
5
4
3
2
1
0
P = 0.03
P = 0.09
Currently marketed
New
Front
Rear

Suppose we get the outcome in Figure 10.10 instead. We decide that the new tire is better than the currently marketed one in both positions, but the difference is greatest when the tire is at the front. Wrong. Here is why.

- Again, we are doing two tests instead of one.
- Although P is lower for the front than the rear, this does not mean that the difference in the differences is meaningful. P is not an indicator of effect size (Chapter 6). While P is influenced by effect size, P is also influenced by sample size and variance. The lower P-value for the front may have been due to larger samples sizes, or less scatter, than for the rear.

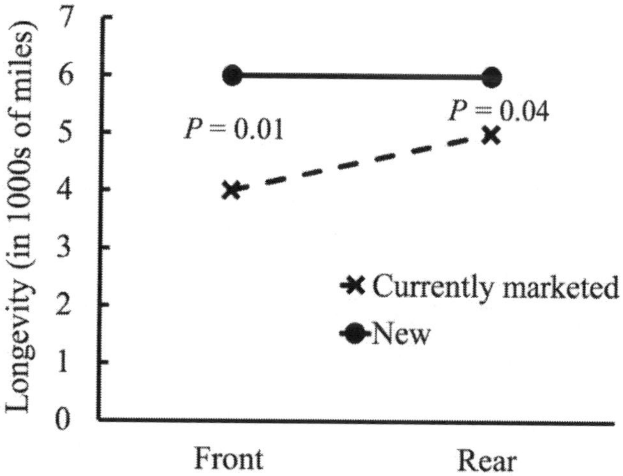

Longevity (in 1000s of miles)
7
6
5
4
3
2
1
0
P = 0.01
P = 0.04
Currently marketed
New
Front
Rear

Figure 10.10 More invalid results of performing two tests instead of a single test for an interaction.

Suppose we do four *t* tests instead of two and get the results in Figure 10.11. Now the familywise error rate is 0.19.[1] People do this thinking that they are teasing apart their data. Actually, they are maximizing their chances of drawing an incorrect conclusion, by performing four tests instead of one.

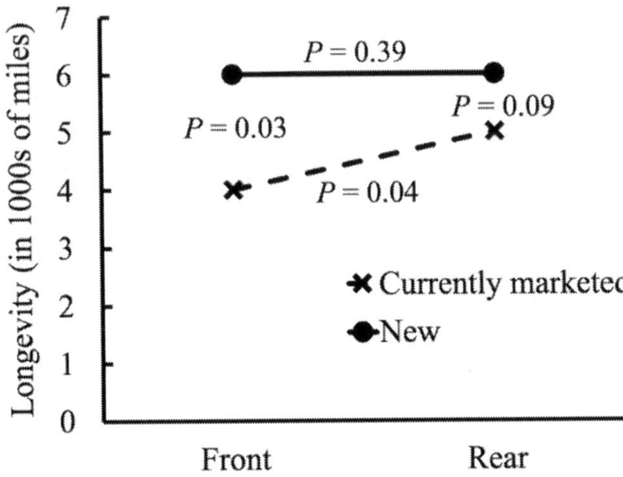

Figure 10.11 Results of performing four tests when only one is justified—a test for an interaction.

10.3 We Have a Main Effect: Now What?

A main effect of one variable can be followed up with post hoc tests (Section 9.3), but we have to account for the other variable. Imagine that we are comparing three types of tires; A, B, and C; and we want to know how long they last, and whether their position matters, front or back. The design is a 2 × 3 (Figure 10.12).

Figure 10.12 A matrix used to illustrate how post hoc tests can be performed to follow up on a main effect.

	Tire A	Tire B	Tire C
Front	$n_{f,a} = 10$ $\bar{x}_{f,a} = 42{,}000$	$n_{f,b} = 10$ $\bar{x}_{f,b} = 52{,}000$	$n_{f,c} = 10$ $\bar{x}_{f,c} = 58{,}000$
Rear	$n_{r,a} = 10$ $\bar{x}_{r,a} = 40{,}000$	$n_{r,b} = 10$ $\bar{x}_{r,b} = 50{,}000$	$n_{r,c} = 10$ $\bar{x}_{r,c} = 55{,}000$

There is a main effect of tire, and we want to know which directional differences we can trust as meaningful. The next step is post hoc tests, like Tukey's or the Tukey-Kramer test (Section 9.3), but modified for a factorial design. Suppose we want to compare tire B with tire C. We calculate standard error as follows.

[1] For why it is not 20%, see J. Zar. 2010. Biostatistical Analysis, 5th ed. Prentice Hall, p. 189,190.

$$SE = \sqrt{\left(\frac{error\ MS}{2}\right)\left(\frac{1}{n_1} + \frac{1}{n_2}\right)}$$

$$SE = \sqrt{\left(\frac{error\ MS}{2}\right)\left(\frac{1}{20} + \frac{1}{20}\right)}$$

Note that the sample sizes, *n*, are from both front and rear combined. They are **marginal sample sizes** (e.g. Figure 10.1). Each level within the factor *tire* is functioning as a single sample, undivided by the other independent variable, *position*. So, for tire B, the sample size is 20, and the same is true for tire C.

Next, we calculate *q* as follows.

$$q = \frac{|\bar{x}_B - \bar{x}_C|}{SE}$$

$$q = \frac{|51{,}000 - 56{,}500|}{SE}$$

For the average longevity of tire B, we average the two cell means to calculate a **marginal mean**. In the front, tire B lasted an average of 52,000 mi. At the rear it was 50,000 mi. We use the average those cells means to calculate a marginal mean of 51,000 mi. We do the same for tire C. From here on, we proceed as if following up on a 1-way ANOVA (Section 9.3).

If we have our favorite (or loathed) statistics package perform Tukey tests, to follow up on main effects, we should check the output's sample sizes and means, to ensure that they are marginal.

10.4 $P \leq \alpha$ For an Interaction: Things to Consider

If $P < \alpha$ for an interaction, there are things to consider before proceeding with more tests.

- To minimize familywise error (Section 9.1), we should try to do as few tests as possible, and they should be specialized post hoc tests when possible.
- In some cases, it will not be possible to control familywise error. We must remember the 5% risk and the fact this should not be a sanctification process (Section 2.6). An unexpected result of $P \leq \alpha$ should be treated with skepticism.
- It may be best to plot the data, with cell means and error bars (Chapter 14), and evaluate them with our intelligence and intuition, rather than to do a lot of additional

tests. In other words, finding an interaction might mark the time to stop testing null hypotheses and accept that we can only extract so much information from a single set of data.

10.5 We Have an Interaction and We Want to Keep Testing Nulls

Suppose we have the results in Figure 10.13, and $P \leq \alpha$ for the interaction. We could do three *t* tests to compare front with back for the three types of tires. Familywise error would be 0.14.[1] The other option is two 1-way ANOVAs, one to compare all three types of tire at the front and the other all three at the rear.[2] Familywise error would be 0.10. Two 1-way ANOVAs is the better option. It gives us fewer opportunities to draw incorrect conclusions when we should not.

Figure 10.13 A matrix that illustrates how post hoc tests can be performed to follow up on an interaction.

	Tire A	Tire B	Tire C
Front	$n_{f,a} = 10$ $\bar{x}_{f,a} = 40{,}000$	$n_{f,b} = 10$ $\bar{x}_{f,b} = 42{,}000$	$n_{f,c} = 10$ $\bar{x}_{f,c} = 43{,}000$
Rear	$n_{r,a} = 10$ $\bar{x}_{r,a} = 40{,}000$	$n_{r,b} = 10$ $\bar{x}_{r,b} = 45{,}000$	$n_{r,c} = 10$ $\bar{x}_{r,c} = 50{,}000$

Suppose we do two 1-way ANOVAs and, for the rear, $P > \alpha$. We stop there for the rear. For the front, $P \leq \alpha$, so it is time to do more tests. They should be post hoc tests designed to limit familywise error. They should be tests like Tukey tests.

Sometimes it is not as simple as choosing the method with the fewest comparisons. Imagine that we have a drug that we hope will improve memory. We want to generalize our findings beyond a single mouse strain, so we choose three strains of mice and two strains of rats. For each strain, we assign some rodents to the control group and others to receive the drug. Each rodent is given a memory task, and we record its score. We find that $P \leq \alpha$ for the interaction. We could follow up with two 1-way ANOVAs, one to compare the five control groups to each other and the other to compare the five treatment groups to each other. The other option is to compare *control* to *treatment* for each of the five strains. The former would carry the lower familywise error rate, since it entails two tests instead of five, but what would we make of the results? Only the latter would involve direct comparisons of treatment groups to controls. When following up on an interaction, **it is important to balance the need to maintain low**

[1] For why it is not 15%, see J. Zar. 2010. Biostatistical Analysis, 5th ed. Prentice Hall, p. 189,190.

[2] Or two sets of Tukey tests without the ANOVAs first, but see Section 9.3.

familywise error with the need to perform the most informative tests. This must be done on a case-by-case basis.

10.6 Designs with More Than Two Independent Variables

We can perform ANOVAs with any number of independent variables. For example, we might want to look at how long two brands of tire last in the front and rear positions of two different kinds of cars (one big car, one small). Since there are three independent variables, results would be analyzed with a 3-way ANOVA. Two problems arise. One is that a 3-way ANOVA tests nine null hypotheses—three for main effects, three for the three pairwise interactions, and one for the three-way interaction. If we set α to 0.05, then over a lifetime of performing 3-way ANOVAs, we will wrongly exclude chance for one of those hypotheses 63% of the time.[1] Since these are separate nulls, there is no way to reduce that number (*cf.* Section 9.1). Follow up tests might include three 2-way ANOVAs, each of which could lead to multiple 1-way ANOVAs and finally post hoc tests. We could have quite a few cases in which we wrongly exclude chance. The other issue is communicating the results of such an exhaustive analysis. A reader is likely to be overwhelmed.

One way to address these problems is to **ignore results that are not interesting**. Suppose we have three independent variables—*car*, *tire*, and *position*—and we find an interaction between *car* and *position*. We ignore the result. What we want to know about is the tires.

Another solution is to **be guided by effect size** (Chapter 6). Imagine we get the results in Figure 10.14. Suppose there is a main effect of *car* (large v. small), a main effect of *tire*, and an interaction of *tire* and *position*. *Car* is unimportant, so we disregard it. What about the main effect of *tire*, and the interaction of *tire* and *position*? We plot the main effect and the interaction to assess their relative importance, i.e., their relative effect sizes. To plot the main effect, we average the cell means for each column to create marginal means. The average for the first column is 29,000 mi. The other two marginal means are 42,500 and 47,500 mi (Figure 10.15).

Next, we plot the interaction. For *front/tire A*, we average the results for the two types of cars. For large cars, it is 30,000 mi, for small it is 28,000 mi. So, for *front/tire A*, we plot 29,000 mi. We do the same for the other five combinations, averaging the results for the two car sizes. The resulting plot shows that the interaction is relatively unimportant (Figure 10.16). For tires B and C, tires last a little longer in the rear than in the front, but the difference is just 3000 mi. In comparison, when we consider the main effect, tire A lasted only 30,000 mi, while tire C went 47,500 mi. The effect size is much

---

[1] For why it is not 20%, see J. Zar. 2010. Biostatistical Analysis, 5th ed. Prentice Hall, p. 189,190.

Figure 10.14 Results of a 3-way factorial design.

Large car	Tire A	Tire B	Tire C
Front	$n_{f,a} = 10$ $\bar{x}_{f,a} = 30{,}000$	$n_{f,b} = 10$ $\bar{x}_{f,b} = 42{,}000$	$n_{f,c} = 10$ $\bar{x}_{f,c} = 47{,}000$
Rear	$n_{r,a} = 10$ $\bar{x}_{r,a} = 30{,}000$	$n_{r,b} = 10$ $\bar{x}_{r,b} = 45{,}000$	$n_{r,c} = 10$ $\bar{x}_{r,c} = 50{,}000$

Small car	Tire A	Tire B	Tire C
Front	$n_{f,a} = 10$ $\bar{x}_{f,a} = 28{,}000$	$n_{f,b} = 10$ $\bar{x}_{f,b} = 40{,}000$	$n_{f,c} = 10$ $\bar{x}_{f,c} = 45{,}000$
Rear	$n_{r,a} = 10$ $\bar{x}_{r,a} = 28{,}000$	$n_{r,b} = 10$ $\bar{x}_{r,b} = 43{,}000$	$n_{r,c} = 10$ $\bar{x}_{r,c} = 48{,}000$

greater for the main effect than the interaction. When communicating our results, it would be best to communicate the main effect. We might mention the interaction, and its relative unimportance, or we might not mention the interaction at all. The latter would help avoid type I errors. Either way, the focus is on the results that are the most important, rather than all of them, which could overwhelm a reader with endless *P*-values, many of which are trivial.

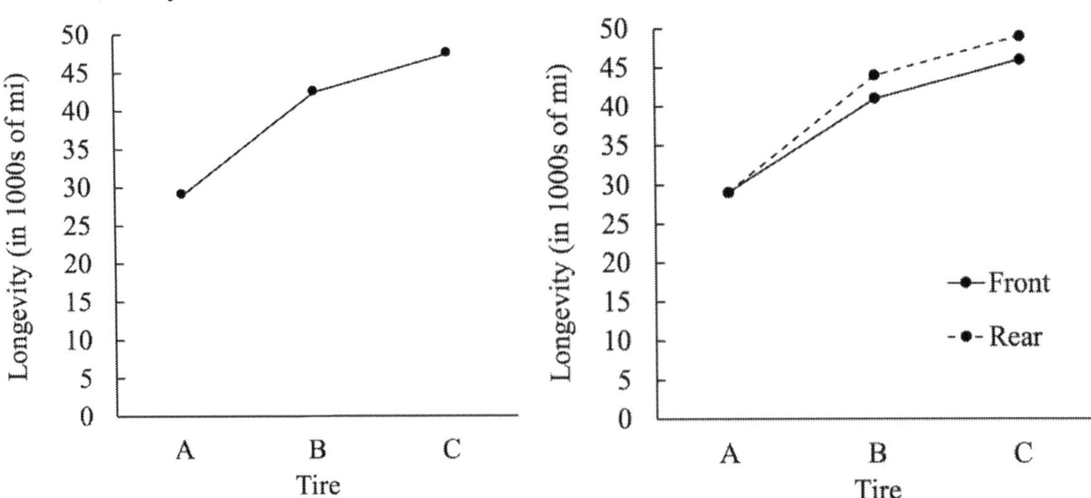

Figure 10.15 Marginal means illustrate a main effect of *tire.*

Figure 10.16 Marginal means illustrate the interaction of *tire* and *position.*

10.7 Use of ANOVA to Reduce Variation and Increase Power

Scatter is the bane of null hypothesis testing. The more scatter in the results, the higher the *P*-value, and the less confident we can be in drawing conclusions (Section 2.7). Sometimes, the variation is partly the result of an independent variable, perhaps a variable we do not care about. By including that variable in the ANOVA, we can remove the variation it causes from the analysis and increase power. Here is how.

Imagine we are assessing the effect of octopamine on the metabolic rate of crickets. We decide to share the burden of the study with a collaborator at a distant institution. We house ten crickets, inject octopamine into five of them, inject a control solution into the others, and measure the crickets' metabolic rates. Our collaborator does the same, but at his or her institution. In dividing things up like that, we have created a variable, *location*, with two levels within it. Since each location will have five treatment animals and five controls, there should be no systematic error caused by *location*. If *location* has an effect, however, it will increase the amount of variation in the two samples, since both locations are represented in the two samples.

Suppose our collaborator performs experiments in a room slightly warmer than ours. Metabolic rates will be higher in our collaborator's crickets. We get the following metabolic rates (Table 10.1).

Table 10.1 Effect of treatment and location on cricket metabolic rate

Our location		Our collaborator's location	
Control	Octopamine	Control	Octopamine
4.34	5.54	6.14	7.34
5.79	6.99	7.59	8.79
4.57	5.77	6.37	7.57
5.43	6.63	7.23	8.43
3.05	4.25	4.85	6.05

If we lump together the results for the two locations and compare *control* with *octopamine* with a 1-way ANOVA,[1] we find that $P = 0.068$. If instead, we perform a 2-way ANOVA, and include *location*, we get the following.

Main effect of location, $P = 0.002$.
Main effect of treatment (control v. octopamine), $P = 0.023$.
Interaction, $P > 0.95$.

---

[1] We would get the same *P*-value by performing a *t* test.

We have no interest in the effect of *location* but, by including it in the analysis, we removed the variation that variable caused in the comparison of *control* with *treatment*. With the 1-way ANOVA, pooled variance (*MS error*, Chapter 8) is 1.91. With the 2-way, it is 1.14. Consequently, $P \leq \alpha$ for the effect of *control* versus *octopamine* only with the 2-way. It may be that we never look at the *P*-values for the interaction or main effect of location. If we look at those *P*-values, we may choose not to report them. If we report them, we point out that they are unimportant.

Chapter 11
Comparing Slopes: ANCOVA

An extremely useful variation on ANOVA is analysis of covariance, **ANCOVA**. With ANCOVA, one of the independent variables is numerical. In chapter 10, all the independent variables are categorical. In addition to being useful in its own right, ANCOVA can be extremely useful to increase a test's power (Section 2.7) and to reduce the effect of a confounding variable.

11.1 Use of ANCOVA to Reduce Variation and Increase Power

Imagine that we are interested in comparing three teaching methods. One teacher is assigned to teach three sections of a course in three different ways. At the end of the term, all students take the same test. This sounds like a great idea, but there is a problem. The students range from very poor to excellent, in terms of their academic abilities. The result will be huge scatter in the end-of-term scores. It is hard to draw conclusions when there is lots of variation. The greater the scatter, the lower the power (Section 2.7).

With ANCOVA, power can be increased by including a numerical variable in the analysis, a numerical variable correlated with whatever causes the scatter. In this case, we could use prior GPA, as it could serve as a proxy for student academic ability. The numerical variable we add is termed the **covariate**. By including GPA as a covariate, we turn an ANOVA into an ANCOVA, and we can correct for the scatter caused by variation in student ability.

As in ANOVA, there are main effects and interactions (Section 10.1). For the following, assume that all differences are meaningful. Suppose we get the results in Figure 11.1. There is a main effect of prior GPA but nothing else. We know that the methods must differ to some degree—the null cannot be true—but we cannot say which of the teaching methods are the best. There is no interaction, which would take the form of differences in slopes.

A less likely outcome is the one in Figure 11.2. There is a main effect of teaching method. Method A is superior to method C. Post hoc tests are justified if we want to compare either of them to method B.

There is no main effect of prior GPA in Figure 11.2, which is why this is an unlikely outcome. Prior GPA is usually a good predictor of future academic performance. We cannot say if a particular teaching method works best for students with a particular prior GPA. There is no interaction. The slopes are indistinguishable from each other.

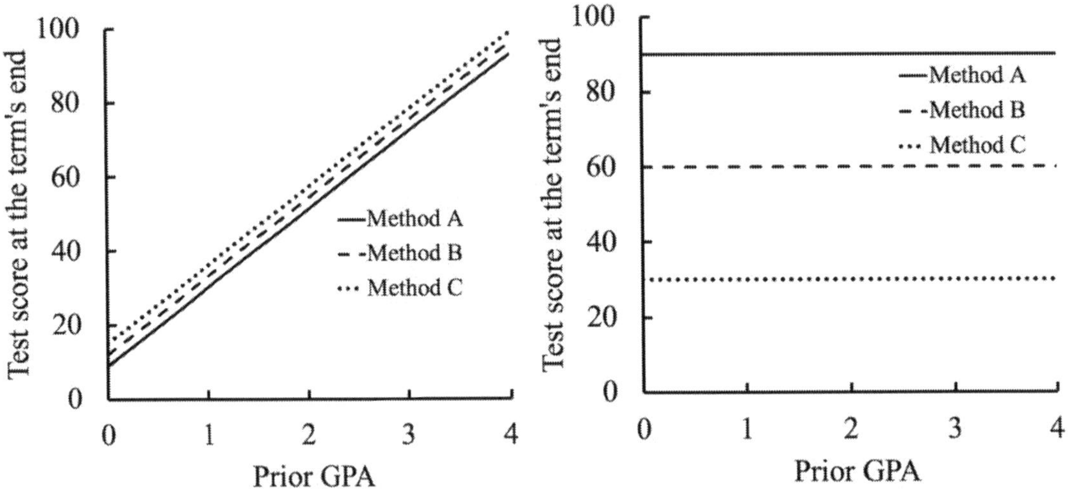

Figure 11.1 A main effect of prior GPA.

Figure 11.2 A main effect of teaching method.

Suppose we get the outcome in Figure 11.3. We have main effects for both teaching method and prior GPA. Method A works better than method C, and we were able to detect that fact in spite of the huge variation in final scores brought about by the wide range of student abilities. Given any particular students' strength or weakness, Method A works better than Method C. See Section 11.2 for a numerical example, which fleshes out this point. Post hoc tests can be used if we want to know how they compare to Method B.[1]

Figure 11.4 shows an interaction. One method works better than another, but it depends on a student's academic ability. Good students do well regardless of the teaching method. They could probably teach themselves. It is with the poor students that the particular method really matters. This example illustrates how ANCOVA can be useful in its own right, rather than only being used to increase power or control for a confound. **ANCOVA lets us compare slopes**.

[1] These would be post hoc tests of *adjusted means*—adjusted for the covariate.

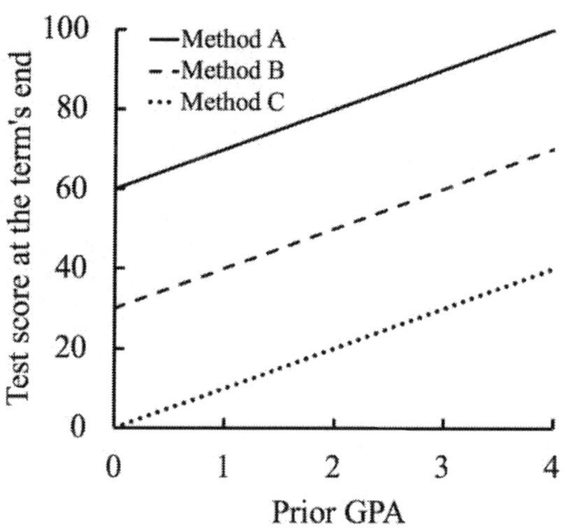

Figure 11.3 Main effects of teaching method and prior GPA.

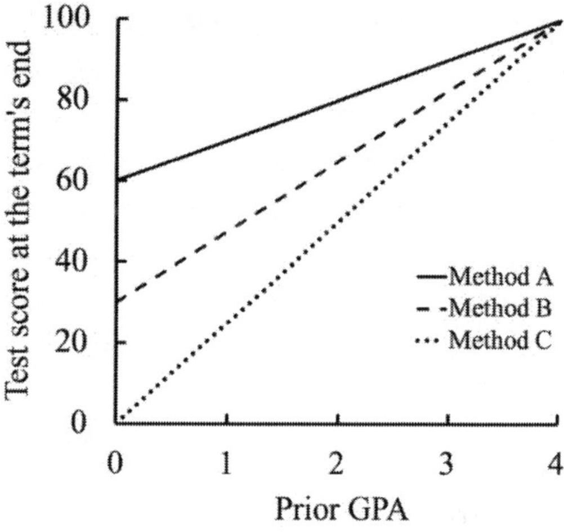

Figure 11.4 In an ANCOVA, an interaction takes the form of meaningful differences in slopes.

11.2 More on the Use of ANCOVA to Reduce Variation and Increase Power

The fact that ANCOVA can be used to increase power is so important that a numerical example is in order. Consider the example above in which we compare the effects of three different teaching methods. Suppose we get the results in Figure 11.5. There is a huge range of final scores, from 49.6 to 100. Also, all three groups overlap from 58.2 to 87.5. If we ignore GPA and perform a 1-way ANOVA,[1] to assess the three teaching methods, we find that $P = 0.544$. We can draw no conclusion.

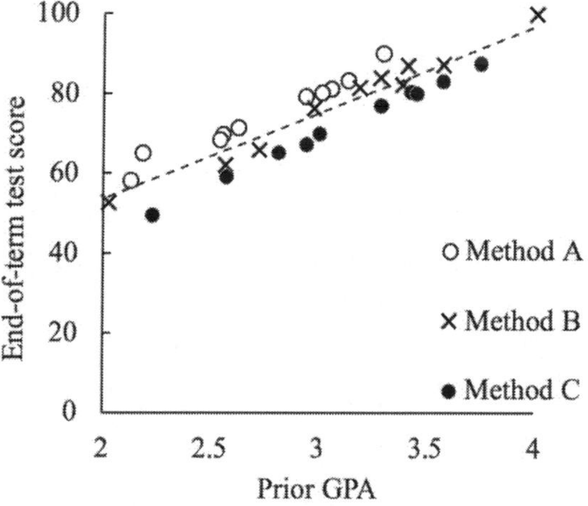

Figure 11.5 Prior GPA accounts for much of the variation in test scores used for the comparison of the three teaching methods, so we have dramatically increased power. The dashed line is the best-fit line for all the data combined.

[1] Or a two-sample t test.

With ANCOVA, we can ask a different question. How do the teaching methods differ given any one student's academic strength or weakness? When we include *prior GPA* as a covariate, we find that $P = 0.039$ for the main effect of teaching method. The variation caused by differences in academic ability is accounted for by *prior GPA*. That variation no longer contaminates the data used to compare the three teaching methods. ANCOVA accomplishes this by comparing the students' scores to the best-fit line, which is based on all data combined. In this case, Method A students are consistently above that line, while Method C students are consistently below it (Figure 11.5).

11.3 Use of ANCOVA to Limit the Effects of a Confound

To illustrate how ANOVA can limit the effect of a confound, consider a case where we want to know if freshmen biology students benefit from attending supplemental instruction (SI) sessions, which are led by upperclassmen. Some students attend SI sessions, and some do not. We want to compare those two groups by looking at their final grades. Since students choose whether to attend SI sessions, the study's test subjects assign themselves to their treatment groups. Subjects should be assigned randomly instead. If subjects assign themselves, the design is said to be **quasi-experimental.**[1] Quasi-experiments are fraught with threats to validity. Maybe it is the highly motivated students that attend SI sessions. If they score better than those who do not attend, it could be because they are more highly motivated, rather than because they attended. The variable *motivation* is a confound. We can limit its impact with ANCOVA.

As in Section 11.1, we need a proxy, this time for the variable *motivation*. We will use high school GPA.

The mean final grade for the students who attended SI sessions is 76.0%. For the students that did not attend, it is 63.0%. If the two samples are compared with a 1-way ANOVA, $P << 0.01$.[2] We would conclude that SI sessions are beneficial.

If, instead, we include high school GPA as a covariate, $P = 0.802$ for the main effect of *attended* versus *did not attend.* When we reduce the effect of the confound, *motivation*, we find that that confound is largely responsible for the low P-value obtained with the 1-way ANOVA. We can draw no conclusion. There must be *some* effect of the SI sessions—the null cannot be true— but, with our P-value, we cannot say whether SI sessions are a good thing or a bad thing.

---

[1] W. Shadish, T. Cook, and D. Campbell. 2002. Experimental and Quasi-experimental Designs for Generalized Causal Inference. Houghton Mifflin Company, p. 14.

[2] Or a two-sample t test.

The ANCOVA results are easy to understand if we look at Figure 11.6. Though student who attended SI sessions scored higher, in general, than those who did not, both groups of students are scattered above and below the best-fit line. Given any particular high school GPA, attending SI sessions has no clear effect on whether a student scores above the best-fit line or below. Compare these results to those in Figure 11.5.

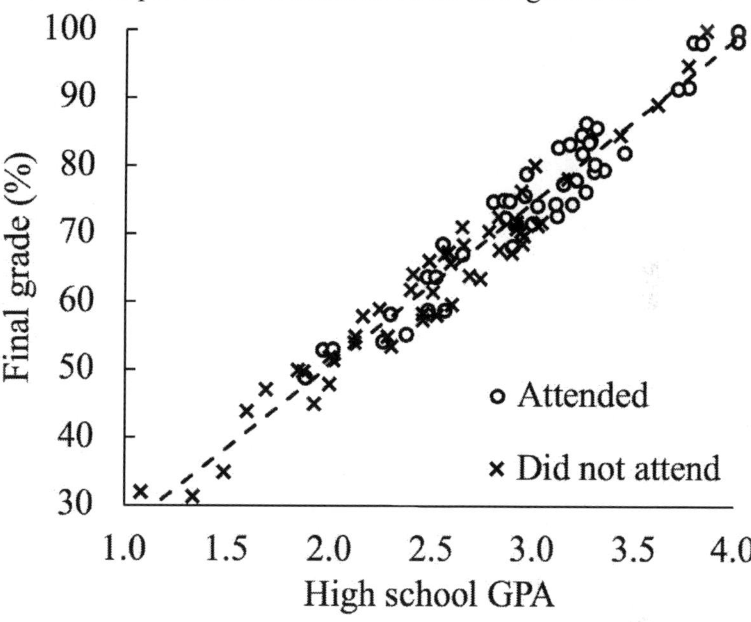

Figure 11.6 High school GPA accounts for the better performance of students who attended supplemental instruction session. The dashed line is the best-fit line for all the data combined.

CHAPTER 12
WHEN DATA DO NOT MEET THE REQUIREMENTS OF *T* TESTS AND ANOVAS

12.1 When Do We Need to Take Action?

When it comes to analyzing numerical data (Section 7.1), we typically use *t* tests or ANOVAs. Both families of tests require that the data have certain properties. The data should be drawn from populations with normal distributions. For our purposes, a normal distribution is a bell-shaped curve that is neither too pointy nor too flat. If we are comparing two or more samples to each other, the populations being sampled must have the same variance. Tests that have these two requirements are termed **parametric**. Typically, we assume that both conditions are met. Parametric tests are **robust,** which means that they give good results even if their requirements are not met, but approximated instead. As long as data do not deviate too far from what is required, they can still be used.

It is when data severely violate the requirements of parametric tests that alternatives are required. To illustrate, consider feeding the single-celled organism *Tetrahymena* on three different concentrations of latex beads and seeing how many beads accumulate after a set period of time. If one concentration is very low, most cells would have very few beads, while a few cells would have many. The result is a **floor effect**. Values cannot fall below zero, so they will be clumped up close to it and skewed upward, away from it. Ivlev's forage index is subject to a floor effect, since forage index ranges from -1 to infinity, but our data in Example 2.1 are not clustered near -1, so it was safe to use a single sample *t* test to analyze them. The opposite of a floor effect can also take place. If bead concentrations are very high, many cells will eliminate beads as rapidly as the cells consume them (through a protist's equivalent of defecation). At that point, most cells would contain the maximum number of beads. The result would be a **ceiling effect**. Data would be clumped close to that maximum number and skewed downward, away from that maximum number. At an intermediate concentration of beads, there might be a normal distribution.

There are two options when data severely violate the requirements of parametric tests. One is to convert the data to values that meet the tests' requirements. Such a conversion is referred to as a **transformation**. The other option is to use a test that does not require normal distributions or equal variances. Such tests are said to be **nonparametric**.

12.2 Floor Effects and the Square Root Transformation

A floor effect can take place if there is a minimum value that data cannot fall below. Most often, that minimum value is zero, but it could be something else, like -1 for Ivlev's forage index. Consider Figure 12.1. One sample is close to zero and skewed away from it. The distribution is not normal. The other sample is normally distributed, but it shows more variation, since it is not forced against a minimum value. When a floor effect is present, we can perform a square root transformation. The formula is as follows.[1]

$$X' = \sqrt{X + 0.5}$$

Each replicate is indicated by X and each transformed value by X'. The effect is to make all values smaller but, the greater the initial value, the farther down it gets moved. When the data in Figure 12.1 are treated with a square root transformation, they appear as shown in Figure 12.2. Both distributions are close to normal, and both have about the same amount of variation within them. The data are now suitable for a t test or ANOVA.

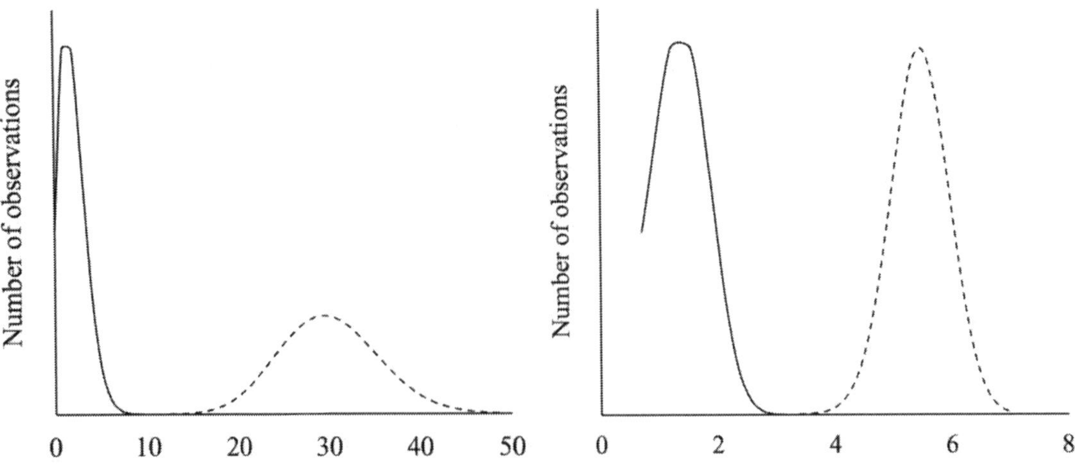

Figure 12.1 A floor effect.

Figure 12.2 Data shown in Figure 12.1 but after a square root transformation.

---

[1] For other more elaborate alternatives, see J. Zar. 2010. Biostatistical Analysis, 5th ed. Prentice Hall, p. 291.

For a square root transformation to work, the minimum value must be zero. Suppose we are working with Ivlev's forage index, and the minimum is -1. We add 1 to all our replicates and perform the transformation.

12.3 Floor and Ceiling Effects and the Arcsine Transformation

Data often have both a minimum value they cannot fall below and a maximum they cannot exceed. Most often, these are 0 and 1, or 0% and 100%. If data fall close to those extremes, they are compressed, and skewed away from those extremes. Consider the example in Figure 12.3.

Of the three samples, only the one with a mean of 0.3 appears to be drawn from a normally distributed population. That same sample also has more variation within it than the others. The solution is an **arcsine transformation**. The formula is as follows.

$$p' = arcsin\sqrt{p}$$

Each replicate is indicated by *p*, expressed on a scale of 0 to 1. Data should be converted to that scale if necessary, e.g., percentages should be divided by 100. We take the square root of each datum and then the arcsine of the result. Transformed values are indicated by *p*'.

The arcsine transformation moves data away from 0 and 1 towards 0.5. The closer the values are to 0 or 1, the farther they are moved toward the middle. As a result, the data in Figure 12.3 are transformed into the data in Figure 12.4. The floor and ceiling effects are nearly gone. The distributions are close to normal, and they have about the same amount of variation within them.

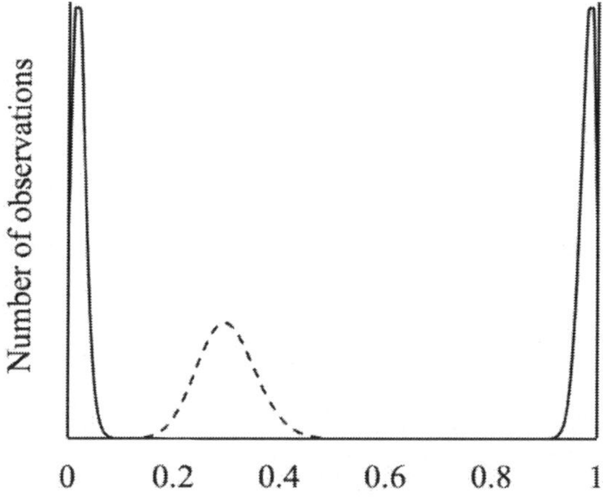

Figure 12.3 A floor and ceiling effect. Data cannot fall below zero or exceed 1.

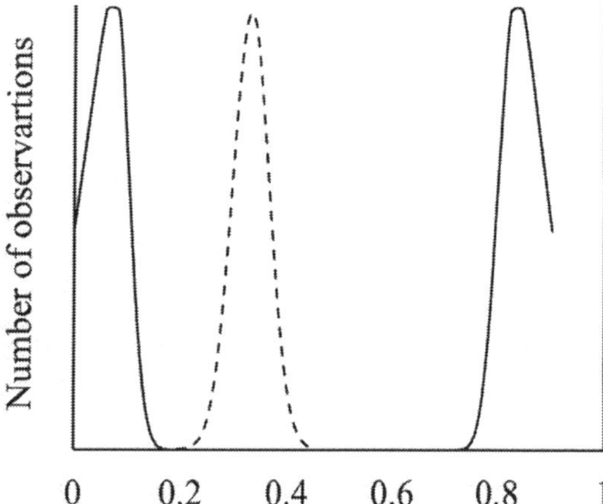

Figure 12.4 Data shown in Figure 12.3 but after an arcsine transformation.

Microsoft Excel's formula for the arcsine transformation is as follows.

=(180/PI())*(ASIN(SQRT(reference cell)))

The term *(180/PI())* ensures that the outcome falls into the range of 0 to 0.9, rather than being in radians.

12.4 Not as Simple as a Floor or Ceiling Effect—the Rank Transformation

Sometimes there are data that violate the requirements for *t* tests and ANOVAs, but the situation is not as simple as a floor or ceiling effect. Often, the solution is to transform the data to their ranks[1] and analyze the ranks with a *t* test or ANOVA. Table 12.1 illustrates how data should be ranked.

Table 12.1 Raw and ranked data. Boxes show an example of how tied ranks should be handled. Ranking failed to solve the problem of different variances.

	Raw data			Ranked data		
	Sample A	Sample B	Sample C	Sample A	Sample B	Sample C
	5.3	9.5	11.5	7	21	24
	5.9	6.9	8.7	8	12	17.5
	6.7	3.7	8.0	11	3	13
	2.9	6.2	8.4	2	9.5	15
	4.3	9.2	8.5	4	20	16
	4.4	8.7	9.1	5.5	17.5	19
	4.4	1.2	9.6	5.5	1	22
	6.2	8.1	11.1	9.5	14	23
Mean	6.6	12.3	18.7	4.5	12.5	20.5
Variance	1.6	8.5	1.7	8.6	55.3	16.1

Note how **tied ranks** should be handled. In sample A raw data, the value 4.4 appears twice (shown boxed). If those two values were sequential instead, they would be ranked 5 and 6. Both of the two 4.4's are assigned the rank of 5.5, since that is the average of 5 and 6. Other cases involved assigning the ranks of 9.5 and 17.5. If software applications are used to rank data, the output should be checked to ensure ensure that tied ranks are handled correctly.

[1] Ranks are considered ordinal data (cf. Section 7.1)

Why would the raw data need to be ranked in the first place? The problem is **large differences in variance**. Sample B's variance is 5.3 times that of sample A. What effect did ranking have on those differences in variance? In this case, ranking made the problem worse. For the ranked data, sample B's variance is 6.4 times that of sample A. This illustrates the fact that **ranking does not always solve the problem**. When raw data have large differences in variance, we need to determine variance for the ranked data, to ensure the problem has been solved.

The reason ranking did not help in Example 12.1 is that there was too much overlap among the three groups. Sample B's mean is almost twice that of sample A, but sample B included the lowest ranking datum as well as the third lowest. Suppose they were spread out more. We could get results like those shown in Example 12.2.

Table 12.2 A second set of raw and ranked data. In this case, ranking solved the problem of different variances.

	Raw data			Ranked data		
	Sample A	Sample B	Sample C	Sample A	Sample B	Sample C
	5.3	24.5	36.5	5	16	24
	5.9	21.9	33.7	6	12	20
	6.7	18.7	33.0	8	10	17
	2.9	21.2	33.4	1	11	18
	4.3	24.2	33.5	2	15	19
	4.4	23.7	34.1	3	14	21
	4.4	16.2	34.6	4	9	22
	6.2	23.1	36.1	7	13	23
Mean	5.0	21.7	34.3	4.5	12.5	20.5
Variance	1.6	8.5	1.7	6.0	6.0	6.0

For the raw data, the variances are the same as in the earlier example, but the means are much farther apart. The first eight ranks are in sample A, the second in B, and the third in C. With no overlap, the variances of the ranked data are identical to each other. Ranking *can* render data appropriate for a t test or ANOVA, but it is important to calculate variance for the ranked data to ensure that the problem has been solved.

12.5 Making ANOVA Sensitive to Differences in Proportion—the Log Transformation

You would think that a log transformation would have an effect similar to a square root transformation. Both would transform large values to smaller ones and, in both, the larger the value the more it is reduced by the transformation. A log transformation, however, has another effect. It makes ANOVA sensitive to differences in proportion, rather than absolute differences.

To illustrate, imagine infecting two groups of mice with a deadly agent, giving each group different treatments for the disease, and determining how many mice survive for various lengths of time. We might get results like those shown in Figure 12.5.

Figure 12.5 Raw data showing the survival of two groups of mice that received different treatments. Best fit lines are based on exponential decay.

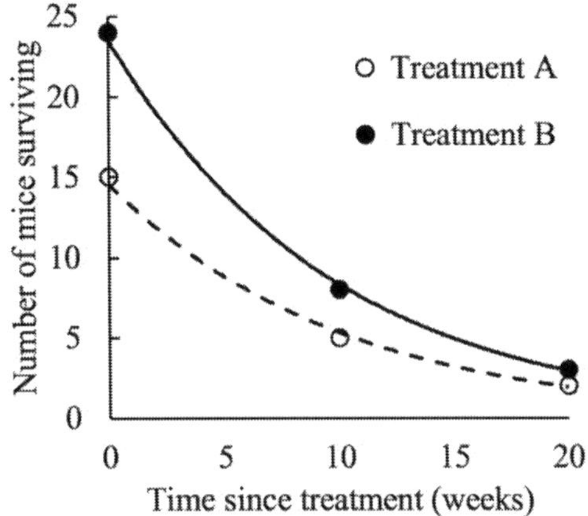

An ANOVA might indicate an interaction (Chapter 10)—at time zero, there are nine more mice in group A than B while, 20 weeks later, the difference is a single mouse. This is not helpful information. We want to know if the treatments differ in how they affect the *proportion* of mice surviving various lengths of time. We could convert the results to percentage of initial, but we could also perform a log transformation on the number of mice and analyze those transformed values. The formula is as follows.

$$X' = \log(X + 1)$$

Each replicate is indicated by X and each transformed value by X'. We would get the results shown in Figure 12.6.

The lack of an interaction indicates that we can draw no conclusion regarding the effect of treatment A versus treatment B on the proportion of mice that survive different lengths of time. Conversely, if there were an interaction, we would be able to determine which treatment led to the greatest proportion surviving, as its line would have the shallower slope. It is also helpful to think in terms of half-life. With no interaction, the half-lives are indistinguishable. If there were an interaction, the shallower slope would correspond to the longer half-life.

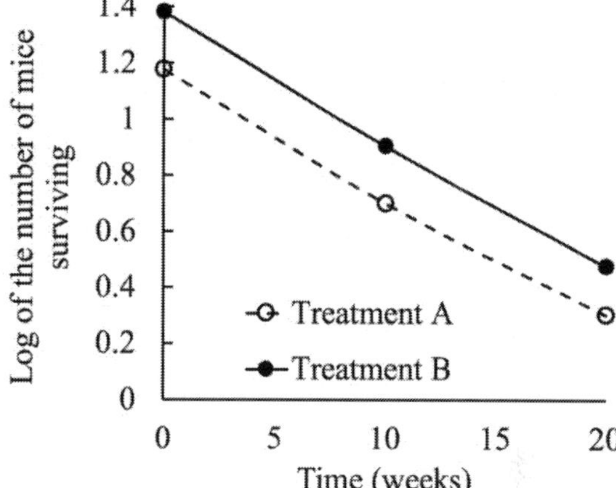

Figure 12.6 The data in Figure 12.5 after a log transformation.

12.6 Nonparametric Tests

Instead of transforming data to render them appropriate for a *t* test or ANOVA, another option is to use a nonparametric test. These tests do not require normal distributions or equal variances. Unfortunately, there is a limited number of such tests available. Some commonly used ones are shown in Table 12.1. If there is a nonparametric test suitable, is it better to use it, or should we transform our data and use a *t* test or ANOVA? In general, parametric tests have a little more power than their nonparametric equivalents, but I would opt for the nonparametric test. Readers are more likely to be familiar with nonparametric tests than with transformations. In fact, some will erroneously think that data were transformed as a way of "massaging" them, to achieve a low *P*-value.

Parametric tests typically require a rank transformation. Tied ranks should be handled as described in Section 12.4.

Table 12.1 Some commonly used nonparametric tests and their parametric equivalents

Nonparametric	Parametric
Mann-Whitney	Two-sample *t*
Wilcoxon signed-ranks, aka, paired samples	Paired *t*
Kruskal-wallis	1-way ANOVA
Friedman's	1-way ANOVA for repeated measures
Nemenji	Tukey post hoc
Spearman rank correlation	*t* test for a slope of zero

CHAPTER 13
REDUCING VARIATION AND INCREASING POWER BY HAVING SUBJECTS SERVE AS THEIR OWN CONTROLS

13.1 The Simple Principle Behind the Mathematics

With a firm understanding of null hypothesis testing, we can use various strategies to increase power. Recall that power refers to a test's ability to detect a directional difference that is not caused by sampling error (Section 2.7). Power is like a test's sensitivity. If a directional difference is meaningful, we do not want the outcome of $P > \alpha$, but that can take place when there is low power. By increasing power, we are more likely to correctly rule out sampling error as the sole cause of a directional difference.

Power can be increased by increasing sample size and by reducing scatter in the results. The latter can be achieved by adding a categorical variable to an ANOVA (Section 10.7), adding a covariate to create an ANCOVA (Chapter 11), and by having subjects serve as their own controls.[1] By letting subjects serve as their own controls, we eliminate variation among subjects from the analysis.

To illustrate, imagine we are interested in the effects of posture on heart rate. We obtain ten subjects and determine the heart rate of each one, both while the subjects are standing and while they are lying down. Suppose we get the results shown in Table 13.1.

If we use a two-sample t test to compare the two sets of heart rates,[2] we find that pooled variance (Chapter 8) is 332 and $P = 0.167$. But what contributes to that pooled variance?

[1] This is a loose use of the word *control*. Narrowly, a control group is a comparison group that does not receive a treatment.

[2] This would not be pseudoreplication (Section 1.4) because each subject provides only one datum per sample.

Table 13.1 Heart rates of 10 subjects in two positions and the subtracted differences of those heart rates.

	Heart rate lying down (beats/min)	Heart rate standing (beats/min)	Difference (beats/min)
	60	94	34
	105	98	-7
	92	105	13
	45	58	13
	80	96	16
	72	75	3
	65	78	13
	90	99	9
	76	77	1
	95	112	17
Mean =	78	89	11
Variance	332 (pooled)		122
$P =$	0.167		0.011

In part, it is the fact that some subjects have higher heart rates than others, regardless of position. For example, heart rates when lying down ranged from 45 to 105 beats/min. There is also variation in how much posture affects the heart rates. If we subtract *lying down* from *standing*, the difference ranges from -7 to 34 beats/min. There is nothing we can do about the varying effects of posture among subjects, but we can eliminate variation caused by the fact that some subjects have higher heart rates than others. We do that by analyzing the subtracted differences with a single-sample t test (Section 7.2), i.e., the right-hand column in Table 13.1 shows the sample we test in the t test. Such a test is usually called a **paired *t* test**, but it is simply a single sample t test on differences.

The result of the paired t test is that variance is reduced to 122 and $P = 0.011$. We have dramatically increased power by making the comparison on a per subject basis. Each subject's heart rate while standing is compared to the same subject's heart rate lying down. Thus, in a sense, each subject serves as his or her own control, and we are making a **within-subject comparison**. Since we measured heart rate twice for every subject, the term ***repeated measures*** is also used. Another term is ***blocking***, which is explained in Section 13.4.

13.2 Repeated Measures ANOVAs

Suppose we have more than two samples. Repeated measures ANOVAs allow for the inclusion of more than two samples, and they are mathematically equivalent to a paired *t* test. If we performed a repeated measures ANOVA with the heart rate data, we would find that $F = t^2$, and we would obtain the same *P*-value as with the paired *t* test.

Suppose we have more than one independent variable. ANOVA allows for any combination of repeated measures and **among-subjects comparisons** (or *between-subjects*), that being the term for *t* tests and ANOVAs that are not paired or repeated measures.

13.3 Post Hoc Tests on Repeated Measures

Post hoc tests, like Tukey's (Section 9.3), can be conducted for repeated measures, but it is important to use the correct form of pooled variance in the calculation of standard error. For a repeated measure, it is *remainder MS*, rather than *error MS*. *Remainder MS* is the error term in which variance has been reduced, so *remainder MS* should be used to calculate standard error for the Tukey test. Otherwise, we lose the power we gained with the repeated measure. This may mean performing the tests by hand, if it is not clear that our favorite (or loathed) statistics package will use the correct error term. To identify *remainder MS*, recall that there is an *F*-ratio for each main effect and interaction. *Remainder MS* is the denominator in the *F*-ratio for the main effect with the repeated measure or any interaction that includes the repeated measure. *Remainder MS* should also be used for creating **error bars** (Section 14.4)

13.4 When Subjects Are Not Organisms

Imagine that we are interested in bird predation on snakes. Do birds avoid snakes that look like venomous species? We identify ten plots of land, all the same size, and place on each one the same combination of four artificial snakes made of Play-Doh. One is yellow, one is blue, one is white, and one is striped to look like a coral snake. Each plot gives us four numbers, the mass ingested by birds for each of the four snakes. Each plot, therefore, is an experimental unit or test subject. The plots may also be referred to as **blocks**, which is why *blocking* is another term for a repeated measures design. In the example of heart rates and posture (Table 13.1), the blocks were the people being studied. Each provided two heart rates. In the case of bird predation, each plot provides four measurements and serves as a test subject. So, units or subjects are not always organisms.

13.5 When *Repeated* Does Not Mean Repeated Over Time

The example of bird predation (Section 13.4) also illustrates that *repeated* in *repeated measure* does not always mean repeated over time. It can mean repeated in space, four measures per plot. For another example, consider the dog food preference ANOVA in Chapter 9. We have five dog foods, and we want to know how dogs like them. We assign ten dogs to each food and find out how much they consume per day. Five groups of ten dogs is fifty dogs. Plus, some dogs will be gluttons, while other are sure to show more restraint, and this will create scatter. It would be better to give ten dogs all five foods simultaneously and see how much they eat of each. The measures would not be repeated over time, but repeated in space, as each dog would get five bowels of food at once. Each dog would be a unit, or block, and the repeated measures ANOVA would remove the variation from dog to dog. Meanwhile, the number of dogs is reduced from fifty to ten, which illustrates another advantage of repeated measures. **Projects are more manageable**.

13.6 Pretest/Posttest Designs Illustrate the Danger of Measures Repeated Over Time

Imagine we are teachers, and we want to document our students' learning gains. At the beginning of the term, we give them a *pretest* to document their knowledge of the topic. At the end of the term, we give them a posttest. We perform a repeated measures ANOVA, or a single-sample t test on the posttest minus pretest differences. The **results are meaningless**. If scores go up, it could be because the students got better at taking our tests—a **practice effect**. If scores go down, it could be a **fatigue effect**—the students are burned out at the end of the term. Pretest/posttest designs require a control group. In this case, the control group would consist of students that take the pretest and posttest, but do not experience the intervention in between. They would take a different course. Results would be analyzed as described in Section 13.7. Wait, there is more. Students register themselves, when they should be randomly assigned to one group or the other (Sections 1.4 and 11.3). Otherwise, it is time to break out Shadish et al.[1] Learning gains? What learning gains?

This problem with pretest/posttest designs illustrates the need for control groups in general, when measures are repeated over time. In the heart rate example (Section 13.1), a crossover design could be used instead of a separate control group. For half of the

[1] W. Shadish, T. Cook, and D. Campbell. 2002. Experimental and Quasi-experimental Designs for Generalized Causal Inference. Houghton Mifflin Company.

subjects, their heart rates would first be determined when they are standing, and next when they are lying down. For the other half, it would be the other way around.

13.7 ANOVAs Versus *t* Tests and Repeated Measures

Suppose we are interested in a drug that might improve memory. We randomly assign mice to two groups: treatment and control. Each mouse is assigned a memory task and scored. Then, treatment mice get the drug we are investigating, control mice get a control solution, and mice are tested again. The result is a pretest/posttest with control design.

How shall we analyze the results? The design is factorial (Chapter 10), as every combination of *before* versus *after* and *control* versus *treatment* is represented. We could perform a mixed 2-way ANOVA, mixed meaning that *before* versus *after* is treated as a repeated measure, while *control* versus *treatment* is an among-subjects comparison. If $P \leq \alpha$ for the interaction, we could draw conclusions about how a change in memory scores over time is affected by the treatment versus the control. A simpler approach would be to subtract *before* from *after*, post- minus pre-, and compare the two sets of differences with a two-sample *t* test. The result would be that $t^2 = F$, and we would have exactly the same *P*-value as we had for the ANOVA's interaction. This again illustrates the fact that the two types of test are mathematically equivalent.

If a *t* test on differences accomplishes the same as a 2-way ANOVA, which one should be chosen? We should consider that 2-way ANOVAs test for main effects, not just interactions. In some cases, main effects are important. In the example of mouse memory, they are not. There is no sense in lumping together *before* and *after* to compare *control* and *treatment* when before treatment should be no different than before control. There is no sense in lumping together *control* and *treatment* to look at a main effect of *before versus after*. Scores might go up because of a practice effect, or they might go down due to fatigue (Section 13.6), but we do not care. In cases where main effects are important, then ANOVA is the way to go. In the case of mouse memory, it does not matter whether a *t* test on differences or a 2-way is used.

Chapter 14
What Do Those Error Bars Mean?

14.1 Confidence Intervals

What do those error bars mean? In a broad sense, error bars can indicate a number of quantities. They may indicate range, quartiles, or standard deviation (Section 7.2.1), in which case they represent the amount of scatter in the data. Because range, quartiles, or standard deviation are not influenced by sample size, those quantities cannot be used to test null hypotheses in our heads. Often, however, error bars indicate confidence intervals (or limits), which can be used—under certain circumstances, cautiously—to test null hypotheses, if *P*-values are not provided.

To understand **confidence intervals**, recall that a sample mean is an estimate of a population mean. How confident are we that the sample mean is a good estimate? It depends on the amount of variation in the data: the greater the variation, the less confident we can be. But it also depends on the sample size. The larger the sample size, the more representative the sample is of the population. By incorporating sample size, along with variation, into confidence intervals, we can use confidence intervals to test null hypotheses in our heads—under certain circumstances, cautiously.

What do confidence intervals indicate? Let us start with 95% confidence intervals. Error bars that indicate 95% confidence intervals mean that, if you sample the same population 100 times, the actual population mean should be within the error bars 95 times. In other words, over a lifetime of constructing 95% confidence intervals, and deciding that the population means are within them, we will be wrong 5% of the time. Sound familiar? If we construct 95% confidence intervals around a sample mean, we are performing the mathematical equivalent of a single-sample *t* test (Section 7.2), i.e., confidence intervals indicate the results of null hypothesis tests.

Along with 95%, error bars often show 99% confidence intervals or standard error of the mean, the latter approximating 68% confidence intervals. So, the term *error bar* can be used narrowly, to indicate standard error of the mean, or more broadly. Authors should indicate what their error bars mean—preferably in words, and not with undefined initialisms.

Because of the relation between confidence intervals and null hypothesis tests, there are **misconceptions** regarding confidence intervals that parallel the misconceptions regarding *P* (Chapter 3). Some think that, if we construct 95% confidence intervals around a sample mean, the probability that the population mean lies within the confidence intervals is 95% (or 5% for out of them). Of course, the probability is either zero or one.[1] The population mean is either within the confidence intervals or it is not. Just as bad are statements such as *we can be 95% confident that the population mean is within the confidence intervals*. Statements like that are meaningless.

14.2 Testing Nulls in Our Heads

Error bars are commonly misinterpreted to mean that, if error bars for two samples do not overlap, we can exclude sampling error as the sole cause of a directional difference. In fact, using error bars to test null hypotheses in our heads is far more involved (unless they are used as in Section 14.5). Cumming et al.[2] provide rules of thumb to estimate *P* with error bars, but none are as simple as *if bars do not overlap* . . . Also, those rules cannot be applied if there are more than two groups being compared, for the same reason we cannot use multiple *t* tests in that situation (Section 9.1), and this also means there is no way to gauge the combined effects of multiple independent variables other than by eye (Chapter 10; cf. Chapter 15 in Cumming[3]). When Cumming et al.'s rules can be used, they are helpful—we can download their paper[4] for free—but readers should be provided with *P*-values. The best option, when there are multiple comparisons, is to plot results as described in Section 14.5.

Suppose we want to compare more than two samples, or we want to assess the combined effects of multiple independent variables, and all we have is standard error bars. No *P*-values. Over a lifetime of studying graphs with standard error bars *and P*-values, we can develop some sense of how to interpret standard error bars without *P*-values. Still, readers should be provided with *P*-values.

---

[1] D. Salsberg. 2002. The Lady Tasting Tea. Henry Holt and Company, LLC. p.122.

[2] G. Cumming, F. Fidler, and D. Vaux. 2007. Journal of Cell Biology, 177: 7–11. G. Cumming. 2012. Understanding the New Statistics: Effect Sizes, Confidence Intervals, and Meta-Analysis. Routledge.

[3] G. Cumming. 2012. Understanding the New Statistics: Effect Sizes, Confidence Intervals, and Meta-Analysis. Routledge.

[4] G. Cumming, F. Fidler, D. Vaux. 2007. Journal of Cell Biology, 177: 7–11.

14.3 Plotting Confidence Intervals

Many statistics packages calculate confidence intervals but, if those packages are not available, or we cannot make sense of them, or we need to modify the calculations (Sections 14.4 and 14.5), it is easy to do them by hand. The easiest of all is standard error of the mean. See Section 7.2.2. Plot the mean plus or minus standard error of the mean. We also need standard error to calculate other confidence intervals, but standard error is simply multiplied by the critical value of *t*. The critical value of *t* is the *t* value that corresponds exactly to α. So, consider Example 7.1. The mean forage index is 0.19 and standard error is 0.083. Because $df = 9$ and $\alpha = 0.05$, the corresponding critical value of *t* is 2.262.

Table 7.2 A probability table for *t*

	P (2-tailed)					
DF	0.50	0.20	0.10	0.05	0.01	0.001
7	0.711	1.415	1.895	2.365	3.499	5.408
8	0.706	1.397	1.860	2.306	3.355	5.041
9	0.703	1.383	1.833	2.262	3.250	4.781
10	0.700	1.372	1.812	2.228	3.169	4.587
11	0.697	1.363	1.796	2.201	3.106	4.437

The 95% confidence intervals would be calculated as follows.

$$CI_{0.95} = 0.19 \pm (0.083)(2.262)$$
$$CI_{0.95} = 0.19 \pm 0.19$$

Results are ambiguous. Microsoft Excel returns a 95% confidence limit of 0.162. Combined, we get 0.19 ± 0.162. This outcome excludes zero, which means that $P < 0.05$. We just performed the same single sample *t* test as in Section 7.2, but backwards! Rather than starting with the sample mean and solving for *P,* we started with 0.05 and solved for the sample mean that would give us that *P*-value. Those who think null hypothesis tests should be replaced with confidence intervals should face the fact that they are reverses of each other. I have read that confidence intervals do not force a decision. Neither do *P* values.

In this case, for plotting, the bars should be 0.162 long.

14.4 Error Bars and Repeated Measures

One thing about confidence intervals is that they are calculated based on variance, and it is important to choose the correct term for variance when there are **repeated measures** (Chapter 13). Use of repeated measures reduces variance, so confidence intervals need to be calculated with that variance reduced. For **standard error** of the mean, one way to do it is as follows[1]

$$SE = \sqrt{\frac{remainder\ MS}{n}}$$

in which *remainder MS* is the denominator in the *F*-ratio that pertains to the repeated measure (Section 13.3). For other confidence intervals, they should be calculated as described in Section 14.3, but with standard error calculated based on *remainder MS* rather than *error MS.* If we are unsure whether our favorite (or loathed) statistics package does this, we must do the calculations by hand.

If all samples within a repeated measure have the same size, confidence intervals will be the same for all samples. This is justifiable because a requirement of ANOVA is that the populations have the same variance. The best estimate of that variance would be *remainder MS*, rather than variance calculated for any one sample.

One downside to making all error bars the same size is that it gives the impression that the investigator does not know what he or she is doing. Other ways to calculate standard error for repeated measures are reviewed by Franz and Loftus.[2]

Suppose we are graphing the results of a mixed design (Section 13.2), i.e., a factorial design in which one variable is a repeated measure and the other an among-subjects comparison.[3] Each mean could have two sets of error bars, one showing standard error calculated as usual, the other calculated as above for the repeated measure. The error bars should be explained in the figure's legend, as is always the case.

14.5 Confidence Intervals and Post Hoc Tests

We will pick up with the example in Section 9.3 in which there are results of Tukey tests on differences in dog food preference. A great way to communicate those results, especially if precise *P*-values are unavailable, is with **simultaneous confidence intervals.** Instead of plotting sample means, we plot the difference between every pair of means. Instead of plotting error bars that represent the result of a single-sample *t* test

[1] G. Loftus and M. Masson. 1994. Psychonomic Bulletin, 1(4): 476–490.

[2] V. Franz and G. Loftus. 2012. Psychonomic Bulleting & Review, 19:395–404

[3] See Section 10.6 for issues pertaining to the presence of more than two independent variables.

done backwards (Section 14.1), we plot error bars based on a Tukey test done backwards. This means that the error bars will have been corrected to control familywise error (Section 9.1). On top of that, simultaneous confidence intervals provide unambiguous information about the relation between P and α.

To illustrate, we will compare New Chow to Old Chow (Sections 9.2 and 9.3). The mean for New Chow is 72.8 g/day (Figure 9.2). For Old Chow it is 58.6 g/day. The difference is 14.2 g/day. To plot that difference plus and minus the 95% confidence intervals, we need the value of standard error used in the Tukey test, which is 4.84 (Section 9.3), and the critical value of q, i.e., the value of q that corresponds to α (0.05). There are five groups being compared, and the number of degrees of freedom associated with *error MS* (unless there are repeated measures, see below) is 45 (Section 9.2). My table does not show 45 degrees of freedom, so we go with the closest, lower degrees of freedom it shows, 40. The critical value of q is 4.039. The product of the standard error and the critical value of q is the 95% confidence interval.

$$95\%\ CI = (4.84)(4.039)$$
$$95\%\ CI = 19.55$$

To compute 90% and 99% confidence intervals, we repeat those calculations, but with the use of critical values that correspond to $\alpha = 0.10$ and $\alpha = 0.01$. We plot all those intervals around each difference between the means. Doing so yields Figure 14.1. If zero lies outside of the 95% confidence interval, $P<0.05$. If zero lies outside of the 99% confidence interval, $P<0.01$, and so on. We have practiced wise use (Section 4.2). Readers can see trends and judge for themselves. There are no precise P-values, but

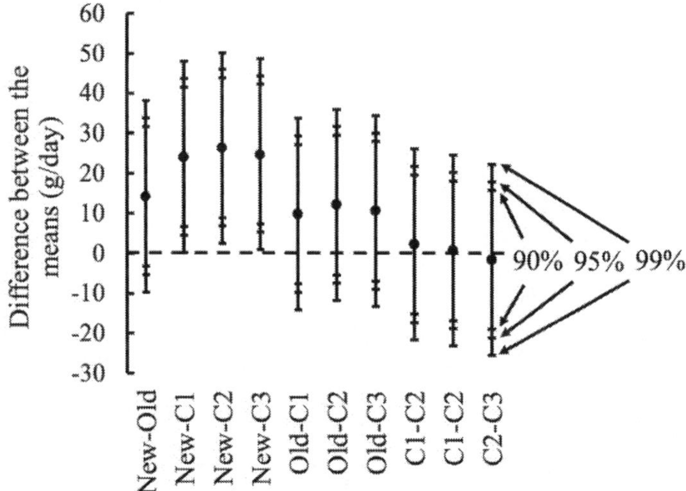

Figure 14.1 Simultaneous confidence intervals around differences provide a lot of information on the outcome of post hoc tests.

it is easy to estimate them. For example, if we compare New Chow to Old Chow, dogs seem to prefer New Chow, but $P > 0.10$. We cannot exclude sampling error as the cause of the preference. Also, the effect size is small. If the purpose of New Chow is to provide better nutrition, it should be marketed. There is no reason to think that dogs will find it less appealing than Old Chow. If the purpose is to provide better flavor, perhaps it should be reformulated.

The same can be done with the results of Dunnett's test (Section 9.3.1), which is used when multiple groups are being compared to a reference, but not to each other. It is simply a matter of using standard error as it is calculated for Dunnett's test, and the critical value of q', which is the test statistic. Variance and degrees of freedom are the same as above for Tukey-based confidence intervals.

When showing the results of Tukey or Dunnett's tests, if the variable is a **repeated measure** (Chapter 13), simultaneous confidence intervals should be calculated with standard error based on *remainder MS* rather than *error MS* (Sections 13.3 and 14.4).

CHAPTER 15
ENLIGHTENED DESIGN OF RESEARCH STUDIES

Advice as to how to design research studies is scattered around this book. Here is that advice all in one place.

- It is important to know the population, and not extrapolate beyond it (Section 14.1). We can think of the population as *that which is sampled*. Pseudoreplication must be avoided. We should study Hurlbert's classic *Pseudoreplication and the design of ecological field experiments.*[1] Pseudoreplication can be subtle, and studying his paper can help us spot it.
- In experiments, subjects should be assigned to treatment groups randomly, not haphazardly (Section 1.3). Another reason to study Hurlbert's classic *Pseudoreplication and the design of ecological field experiments* is that he provides key advice regarding the randomization of other aspects of research studies.
- When studying natural variation, effect size should be reported, even if $P>\alpha$ (Chapter 6). In experiments, we may want to manipulate independent variables to maximize effect size and achieve proof of principle.
- In general, it is best to aim for equal sample sizes, to maximize power (cf. Section 9.3.1).
- When comparing multiple groups to a reference, but not to each other, we should maximize power by following Dunnett's recommendations on sample sizes and using Dunnett's post hoc test to follow up on meaningful results (Section 9.3.1).
- We should maximize power by adding variables to ANOVAs when doing so would help reduce *error MS* (Section 10.6.1).
- We should maximize power by adding covariates to ANOVAs when doing so would help reduce *error MS* (Section 11.1).
- We should maximize power by using repeated measures whenever we can, while controlling for practice and fatigue effects (Chapter 13).

---

[1] S. Hurlbert. 1984. Ecological Monographs, 54(2): 187–211.

ABOUT THE AUTHOR

Frank Corotto earned his bachelor of science in biology at Lafayette College, his master of arts in biology at Boston University, and his doctorate in biological sciences at the University of Missouri–Columbia. Since 1995, he has been at North Georgia College and the University of North Georgia.

Made in the USA
Columbia, SC
10 March 2020